CAMBRIDGE MONOGRAPHS ON PHYSICS

GENERAL EDITORS

M. M. WOOLFSON, D.Sc.
Professor of Theoretical Physics, University of York

J. M. ZIMAN, D.Phil., FRS
Henry Overton Wills Professor of Physics, University of Bristol

Celestial masers

Celestial masers

A. H. COOK, FRS, FRSE
Jacksonian Professor of Natural Philosophy, University of Cambridge

CAMBRIDGE UNIVERSITY PRESS
CAMBRIDGE
LONDON · NEW YORK · MELBOURNE

Published by the Syndics of the Cambridge University Press
The Pitt Building, Trumpington Street, Cambridge CB2 1RP
Bentley House, 200 Euston Road, London NW1 2DB
32 East 57th Street, New York, NY 10022, USA
296 Beaconsfield Parade, Middle Park, Melbourne 3206, Australia

© Cambridge University Press 1977

First published 1977

Printed in Great Britain at the
University Press, Cambridge

Library of Congress cataloguing in publication data
Cook, Alan H.
Celestial masers
(Cambridge monographs in physics)
Bibliography: p. 124   Includes index
1. Masers, celestial   I. Title
QB790.C67   523   76-14028
ISBN 0 521 21344 4

Contents

Preface	*page*	vii
1 Introduction		1
2 Molecular structure and spectra		10
3 Observations of celestial masers		28
4 Theory of amplification by stimulated emission		62
5 Pumping schemes		90
6 Analysis and interpretation of maser radiation		107
Appendix 1 The Born–Oppenheimer approximation and wave functions for rotating molecules		118
Appendix 2 Theories of amplification by tubular and spherical two-level masers		120
References		124
Index		133

Preface

It is over ten years since radiation from maser sources in the galaxy was first detected by the Radio Astronomy Group led by Professor H. Weaver at Berkeley. At first the authors assigned the radiation to 'mysterium' for it seemed unlikely that it could come from a known molecule. However, it soon became clear that the molecule was hydroxyl, long recognized in the night sky, and now maser action is known from water and very probably from other molecules such as CH and SiO. But mysterious the radiations still are. We believe we understand in a general way how they arise, why they are so intense, perhaps why they show strong circular polarization and why they vary in time. Other matters are still very puzzling – why the sources are distributed in space as they are, how the populations are inverted, the details of spectra and so on. The observations therefore pose problems of mathematics – to work out in detail how radiation is transferred in a gas amplifying by stimulated emission, and they pose problems in physics – to identify the mechanisms by which populations become inverted. At the same time, the observations are of great astrophysical interest. Although the strongest evidence of departure from thermodynamic equilibrium, they are not the only ones, and the gas in regions where stars are forming or have just formed, regions from which maser radiation comes, departs in a number of ways from thermodynamic equilibrium. It seems very likely that a better understanding of maser sources may lead to a more detailed knowledge of how gas condenses to stars.

The interest of maser sources is evident; it is perhaps less evident that this is a good time to write a book on them, for observations continue to be published quite frequently, and there are many problems, as indicated, still to solve. So I have not presumed to write a book that will sum up the subject and give a systematic explanation. Rather I have tried to bring out what is not known; no doubt I shall provoke disagreement, and if that disagreement leads to new knowledge and insight, much of my purpose will have been achieved. This is meant to be an interim report and a stimulus for further study. It has fascinated me for ten years and I hope others will find it as interesting.

Preface

I am indebted to many colleagues for help, for discussion, for letting me see their results and telling me of their ideas about them. I must mention Dr P. L. Bender, Professor H. H. Weaver, Dr Nannielou Dieter, Professor R. D. Davies, Dr D. ter Haar, Dr M. Pelling, but there are many others who will I hope excuse their omission. I am grateful to those who have allowed me to use their diagrams in this book, and to my colleagues and students, Dr S. Shafik, Dr M. Salem, M. S. Middleton and J. Viney for their inspiration.

The editor of the series (Professor Woolfson) and Mr M. S. Middleton have removed many blemishes by their careful reading of the typescript.

Cambridge
July 1976

A. H. Cook

1
Introduction

The spectra of radio waves received from most objects in our own and other galaxies are continuous, that is to say, the power does not change abruptly with frequency. By contrast, the characteristic features of the visible spectra of most astronomical objects are absorption or emission confined to rather narrow bands of frequency. The reason for the difference between the radio and visible spectra is that the energies of most transitions between different states of atoms lie in the visible or ultra-violet regions of the spectrum, and radio frequency electromagnetic fields are neither excited by nor excite atomic transitions. Radio frequency fields arise from random motions of electrons (thermal radiation), or from motions of electrons in magnetic fields (synchroton radiation), each of which processes generates a continuous spectrum. There are nonetheless atomic transitions, those between hyperfine levels, with energies corresponding to radio frequency transitions; but, with one important exception, the transitions are too weak to be detected. The exception is the hyperfine transition in the ground state of atomic hydrogen at a frequency of 1421 MHz; although the transition probability is very low (10^{-21} s^{-1}), the numbers of hydrogen atoms in the space between the stars in our own or other galaxies are so great that there is sufficient radiation to detect, as van der Hulst first pointed out. The temperature of interstellar atomic hydrogen is very low (less than 100 K) so that the r.m.s. width of the emission line (supposing it to have the usual Gaussian Doppler profile) is about 6 parts in 10^6 or 8 kHz; mass motions within clouds of hydrogen may however broaden and distort the thermal Doppler line shape, and the movement of a whole cloud along the line of sight will displace the central frequency of the line. It is well known that from observations of the spectrum of the 1421 MHz radiation in different directions the spiral structure of the clouds of hydrogen in the galaxy can be mapped out, the velocities within the clouds can be investigated, and the temperature and density of the atomic hydrogen may be estimated.

As soon as the hydrogen radiation had been detected, the possibility of observing other radio frequency transitions was naturally discussed, and in particular, Shklovskii (1946, 1952, 1953) and Townes (1957) drew up

Introduction

lists of transitions that might be found. Hyperfine transitions in other atoms are unlikely to be detected, for while the transition probabilities are of the same order as that of hydrogen, the numbers of other atoms are very much less. In molecules, however, there are also low frequency transitions with much greater probabilities, so that although the numbers of molecules may be small, the radiation is detectable. The transitions have quantized energies of internal vibration and of rotation as a whole; the frequencies of transitions between vibrational states fall in the infra-red, but transitions between the lower rotational states of some molecules lie in the microwave range. Because the angular momentum of vibration of a diatomic molecule is an integral multiple, J, of $\hbar$, the energy of rotation is equal to $J(J+1)\hbar^2/2I$, where I is the moment of inertia of the molecule. Thus, the greater are the I, the smaller are the energies. The energies of simple hydrides (OH, CH for instance) with small moments of inertia, are too great for the transitions to fall in the microwave or radio range, but the transitions of heavier diatomic molecules (CO for instance) and many polyatomic molecules do lie there. However, a given rotational and vibrational state of a molecule may be split into more than one level by internal interactions and the transitions between such levels may lie in the radio or microwave spectrum. The two examples of importance in astronomy are the so-called Lambda (Λ) Doubling (p. 16) of rotational states of hydroxyl (OH) and the Inversion Spectrum of ammonia. The recently discovered radiation from CH (Rydbeck, Elldér and Irvine, 1973; Turner and Zuckerman, 1974) also comes from lambda-doublet transitions. The probabilities of transitions between states of different rotation are typically between 10^{10} and 10^{15} times as great as those of hyperfine transitions, so that spontaneous emission from a population of molecules in thermal equilibrium may be observed although the numbers of molecules are far far fewer than the numbers of hydrogen atoms.

The transition probabilities of the lambda-doublet transition in hydroxyl are much less than those of typical rotational transitions, but even so, hydroxyl was first detected in 1963 (Weinreb, Barrett, Meeks and Henry, 1963) by the absorption of radiation by the hydroxyl in dense cool clouds of dust and gas. Two years later, however, in 1965 (Weaver, Williams, Dieter, Nannilou and Lum, 1965) emission from hydroxyl molecules was observed. The sources were close to H-II regions (clouds of ionized hydrogen) and the properties of the radiation were, as was recognized independently by a number of authors (Litvak, McWhorter, Meeks and Zieger, 1966; Perkins, Gold and Salpeter, 1966; Cook, 1966; Weinreb and others, 1965) inconsistent with spontaneous emission from a gas in thermal equilibrium. It was soon widely appreciated that amplifi-

Introduction

cation by stimulated emission must be taking place. For some three years, hydroxyl was the only molecule to be found in interstellar space by microwave radiation, but then in 1968, others were detected, either in absorption or by spontaneous emission, so that now some 36 molecules and nearly 200 transitions are known (Rank and others, 1971; Snyder, 1972). Intense stimulated emission is observed from water (Cheung and others, 1969), while methanol, CH and SiO may radiate stimulated emission (Hills, Pankonin and Landecker, 1975) as well as a molecule so far unidentified (Snyder and Buhl, 1974).

Although over eleven years have passed since the first discovery of stimulated emission from hydroxyl, and although some sources have been investigated in detail, the nature of the sources is still obscure in important respects. If stimulated emission is to amplify a beam of radiation then, as in a laser, the net emission, proportional to the number of molecules in the upper state of the two coupled by the transitions, must exceed the net absorption, proportional to the number of molecules in the lower state. The populations are then said to be *inverted*, because in thermal equilibrium the number in the upper state is less than the number in the lower in the ratio of the Boltzmann factor $\exp[-h\nu/kT]$, where ν is the frequency of the transition between the states. If a cloud of gas is amplifying by stimulated emission then it is not in thermal equilibrium. How thermal equilibrium is upset and inverted populations are produced is still unknown. The properties of radiation amplified by stimulated emission have also still to be worked out in detail.

Hydroxyl is rather exceptional among diatomic molecules in that in its ground state the quantum number of the total electronic angular momentum is 1, and that of the total electronic spin is $\frac{1}{2}$. Accordingly, there is an hyperfine interaction with the spin ($\frac{1}{2}$) of the proton and each of the lambda-doublet levels is split into two; the separations of the four levels are shown in figure 1.1, together with the frequencies of the observed transitions between numbers of the lowest rotational state. For convenience, let the levels be labelled 1 to 4 in order of ascending energy.

The observed transitions are then

$4 \to 1$: 1720 MHz

$4 \to 2$: 1667

$3 \to 1$: 1665

$3 \to 2$: 1612

The probabilities of the transitions are not all the same; as explained in chapter 2, they fall into two pairs, the *strong* or main lines at 1667 and

Introduction

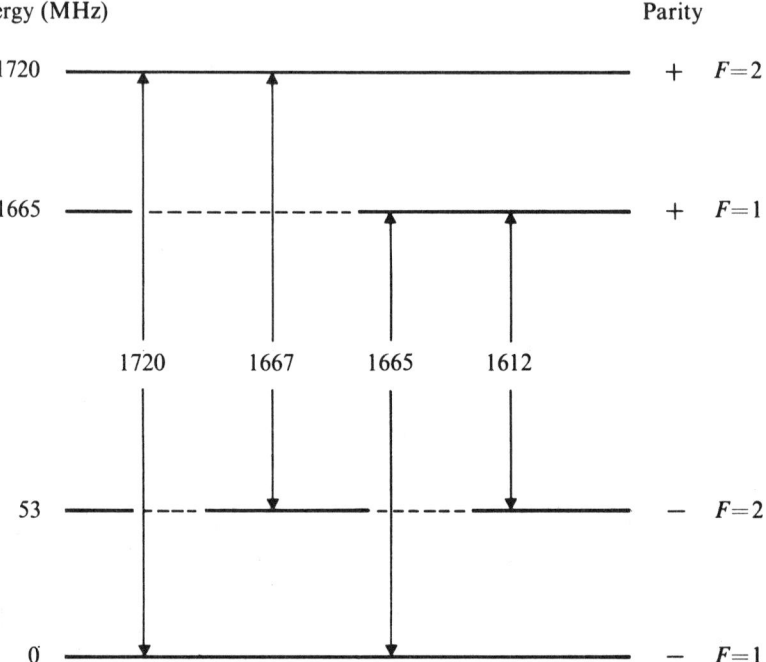

Figure 1.1. The levels of the Λ-doublet set in the ground rotational state of hydroxyl; energies in terms of equivalent frequencies (MHz).

1665 MHz and the *weak* or *satellite* lines at 1720 and 1612 MHz. An optically thin layer of gas (one in which the chance of absorption of a photon is small) would absorb radiation in the approximate ratios:

1720	1667	1665	1612
1	9	5	1

Such a layer would also emit radiation by spontaneous random transitions from upper to lower levels in the same ratios.

If the layer of gas were stationary each line would have a width determined by the temperature of the gas, namely $2(\nu/c)(kT/m)^{\frac{1}{2}}$, where ν is the frequency of the transition, c is the speed of light, k is Boltzmann's constant (1.38×10^{-23} J K^{-1}) and m is the mass of the hydroxyl molecule (2.8×10^{-26} kg). Taking ν to be 1666 MHz, the half width of a line emitted or absorbed by gas at a temperature of 100 K would be 2.4 kHz.

Now suppose that the gas is not stationary but that different parts of it are moving with different mass velocities; absorption and emission will not then occur at a single frequency but over a range of frequencies that

Introduction

differ from those of a gas at rest by vv/c, where v is the mass velocity of a part of the gas. Considering an optically thin cloud of gas, the intensity of emission or absorption at a particular frequency will be proportional only to the optical depth at that frequency, that is, to the mass of gas along the line of sight with the corresponding velocity. Thus, the spectrum of absorption by, or emission from, an optically thin layer of gas should be similar for all four transitions, being determined by the mass velocities as indicated in figure 1.2. The width of each component, as distinct from its position, is determined by the temperature. Spectra of similar form and with amplitudes in the $1:9:5:1$ ratios are almost never observed, whether in absorption or emission. The main feature of absorption spectra is that the intensities of the transitions are much more nearly equal, indicating that the absorbing gas is not optically thin. The mass velocities of the gas derived from the forms of absorption profiles are usually consistent with those deduced from hydrogen 21 cm emission.

The emission spectra are far more complex and cannot be accounted for by spontaneous emission in a gas that is optically thick. In the first place the radiation is exceedingly intense. When a typical source is observed with a single antenna a set of spectra such as that shown in figure 1.3 is obtained. The ordinate of that diagram is the power received as expressed by the aerial temperature (see p. 34) while the abscissa is the frequency expressed as the velocity which would produce a Doppler shift equal to the displacement of the observed frequency from that which would be observed from a source stationary with respect to the Earth. The average equivalent velocity is not zero because sources as a whole will be moving relative to the Earth; the spread of frequencies on the other hand corresponds to the spread of velocities of gas in the source. If the angular extent of the source as seen from the Earth were greater than the angle over which the aerial is sensitive, then the radiation temperature of the source would be equal to the aerial temperature. The radiation temperature is that of a black body radiating with the intensity of the source at the wavelength of the radiation. However, the sources are of very much smaller angular extent than the sensitivity of even the largest single aerial; indeed, interferometric measurements with intercontinental baselines show that the angular diameters can be as small as 10^{-2} arc sec, or even less. The intensity of the radiation is therefore *very* much greater than indicated by aerial temperatures and may correspond to radiation temperatures of the order of 10^{12} to 10^{13} K.

This fact alone shows that the emission cannot come from spontaneous transitions from upper to lower levels, for were that so, the gas would have to be in thermal equilibrium at the radiation temperature; the

Introduction

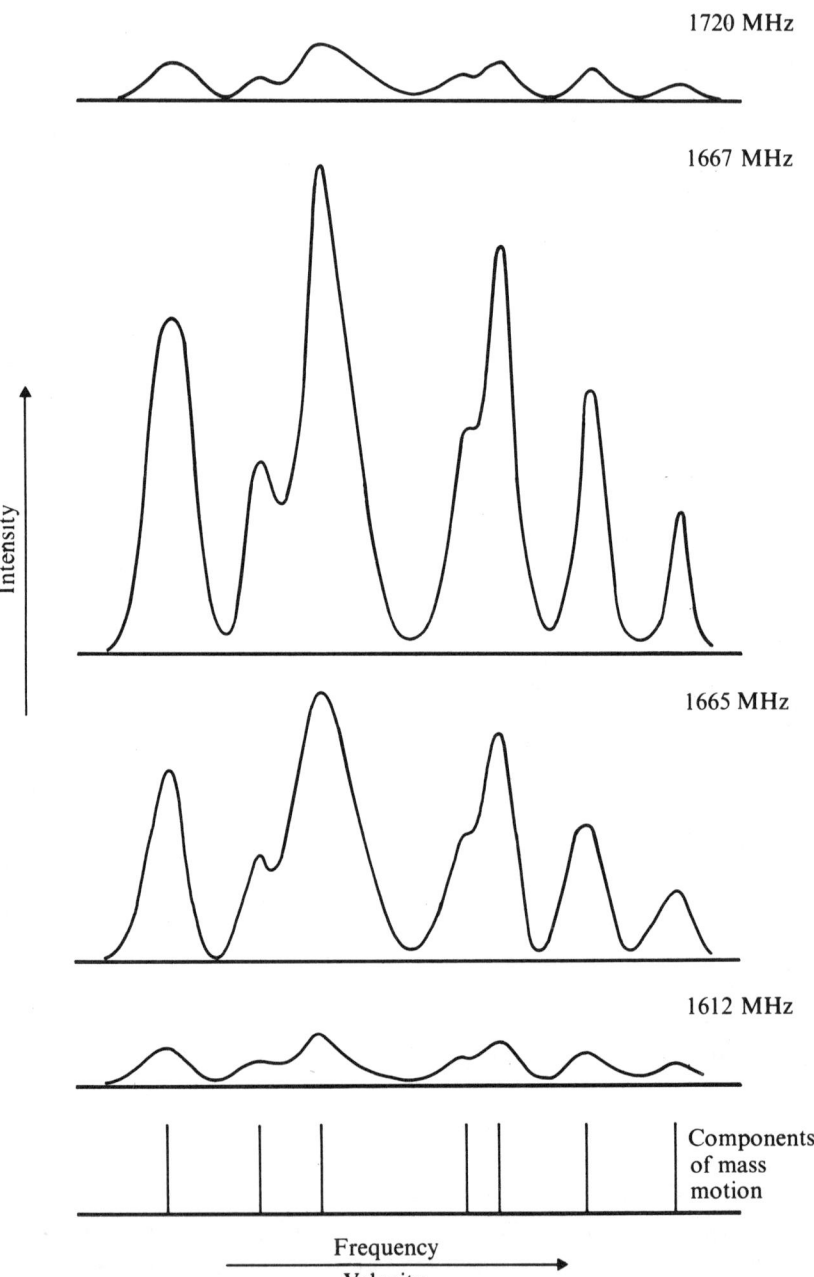

Figure 1.2. Ideal spectrum of spontaneous emission from optically thin volume of hydroxyl with mass motions. Abscissa: velocity (km s^{-1}) equivalent to Doppler shift of frequency. Ordinate: intensity (arbitrary units).

Introduction

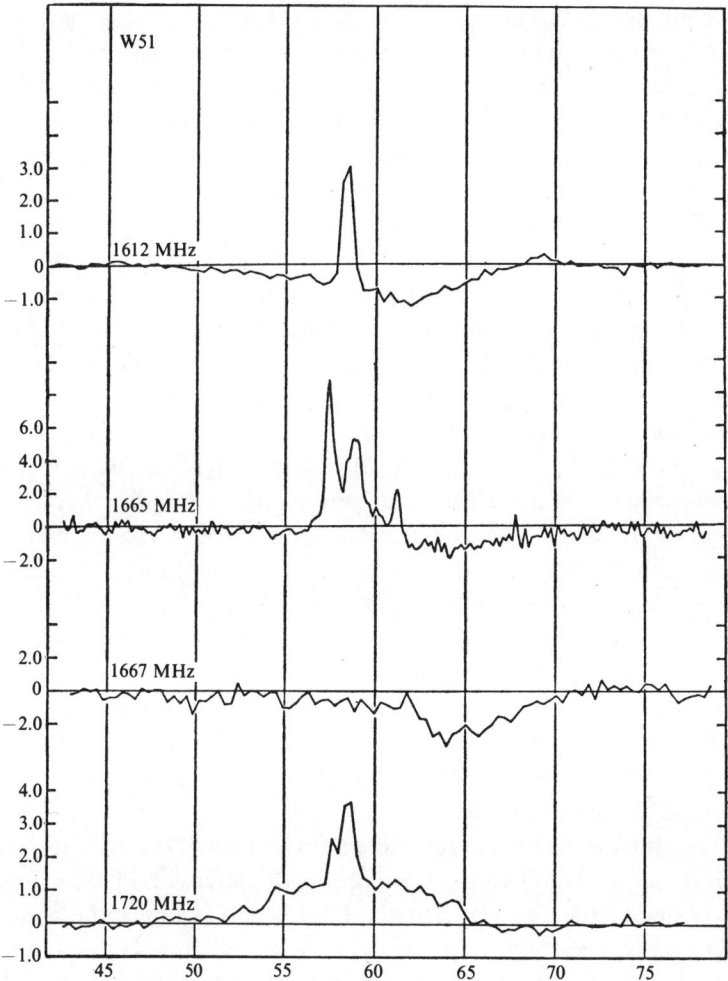

Figure 1.3. Typical emission spectrum of intense radiation from hydroxyl (Weaver, Dieter and Williams. © 1968. The University of Chicago). Abscissa: velocity (km s^{-1}) equivalent to Doppler shift of frequency. Ordinate: aerial temperature (K), proportional to intensity.

random velocities of the molecules would then correspond to a temperature of 10^{12} K or so, and the widths of the emission lines would be very great instead of, in fact, corresponding to a temperature of some 50 to 100 K.

A more detailed examination supports the conclusion that the observed emission cannot come from spontaneous transitions. The spectra clearly do not have similar shapes, so that the shapes cannot be

Introduction

simply due to the Doppler shifts generated by the mass motions of the radiating gas. Indeed, if the intensities at a particular Doppler shift are considered, it will be seen that at some shifts the main line is more intense than the satellite lines whereas at other shifts, a satellite is more intense and indeed, a main line may even show absorption and not emission. This again is inconsistent with spontaneous emission.

Thirdly, the radiation from most sources is strongly polarized, in many instances almost a 100 per cent circularly polarized, both right handed and left handed from the same source at different frequencies. This may not be completely inconsistent with spontaneous emission, but it does make spontaneous emission highly unlikely. Spontaneous radiation is of course polarized if the radiations are in a magnetic field, but the spectrum is quite different from those observed. Spontaneous emission would show symmetrical Zeeman splitting with left and right circularly and linearly polarized components from the same source region; the characteristic of hydroxyl sources is that when interferometric observations are made just one component is seen in any particular direction, and that component is *either* left *or* right circularly polarized.

The fourth feature of radiation from hydroxyl sources is that the intensity varies fairly rapidly in time, characteristic times being of the order of months for hydroxyl sources, certainly very much less than the life times (10^{12} s) of the upper states for spontaneous emission. Once again the implication is that spontaneous emission is most improbable, although not impossible.

All the foregoing features were seen in most of the early observations of radiation from hydroxyl. Subsequently, intense radiation from water was discovered at a wavelength of 1.35 cm (Cheung and others, 1969). The individual sources appear to be even smaller and more intense than the hydroxyl sources, some corresponding to temperatures as high as 10^{15} K. The radiation comes from a transition between two rotational states; with very close hyperfine components there is effectively just the single transition and so there is no question of incompatibility of spectra of different transitions that is so striking a feature of the hydroxyl radiation. Another difference is that the radiation from water is not strongly polarized and when it is, the polarization is linear, not circular. Finally, the intensity of the water radiation may vary much more rapidly than that from hydroxyl, sometimes significantly in a few days.

The distinctive features of the hydroxyl and water radiations are inconsistent with spontaneous emission for if that is the source of the radiation, the intensity, frequency and polarization would be determined solely by the numbers of molecules moving with a given mass velocity and by the magnetic field in the source, and, as has been seen, those parameters alone cannot account for the observations. The alternative is that the radiation comes from stimulated emission. The net stimulated emission from any small volume of gas is proportional to the intensity of the radiation falling on it multiplied by the difference between the numbers of atoms in the upper and lower states. Very much higher intensities can be achieved than in spontaneous emission, and are not limited by thermodynamic equilibrium, while emission from different volumes of the gas is coupled together by the travelling radiation field. The consequent coherence introduces additional variables that may be invoked to account for the observed behaviour of the sources. However, although spontaneous emission is ruled out by the observations, and stimulated emission must be responsible for the radiation, it has not been demonstrated even in principle, let alone in detail, that all the features can be explained by stimulated emission alone.

The ground rotational state of the hydroxyl molecule is not the only one from which stimulated emission comes. Maser radiation is also observed from hydroxyl with higher rotational energy, and pumping processes which may operate for the ground state may not work for one of those higher states.

The first hydroxyl sources to be discovered were associated with H-II regions, volumes of completely ionized atomic hydrogen surrounding hot stars emitting strong ultra-violet radiation. Most sources subsequently found are also associated with H-II regions, but some are associated with infra-red stars or remnants of super-novae or planetary nebulae. Water radiation may be found where there is hydroxyl radiation.

My aim in this book is to describe the principal results of observations of hydroxyl and water sources, to discuss how far they may be accounted for by the theory of amplification by stimulated emission, so far as it is developed at present, and to give a critical summary of pumping processes that have been suggested. I shall try to call attention to problems of astrophysics and molecular physics raised by the observations that await solution. Astrophysicists hope that analysis of the observations will provide information about conditions in the sources which, because of their frequent association with places where stars form, are clearly of great potential interest in astrophysics. Unfortunately, no very definite results can yet be given, and at the same time, problems of the physics of the hydroxyl and water molecules are brought to attention.

2

Molecular structure and spectra

2.1. General remarks

Molecules consist of two or more nuclei surrounded by the cloud of electrons contributed by the constituent atoms. The relative positions of the nuclei are determined by the balance between the repulsive nuclear forces and the attractive forces generated by the electron cloud, in such a way that the total energy is minimized. The forces generated by the electrons need not necessarily be attractive but a molecule cannot be stable unless they are; repulsive forces mean that the constituent atoms fly apart instead of forming a molecule. To a first approximation the relative position of the nuclei may be considered fixed and the nuclei as a group may rotate with respect to any fixed set of axes as a rigid body. Thus the set of nuclei has in general three distinct moments of inertia about three mutually perpendicular axes. The moment of inertia about any axis is

$$\sum_i m_i p_i^2$$

where m_i is the mass of the ith nucleus and p_i is the perpendicular distance of the nucleus from the axis drawn through the centre of mass of the set of nuclei.

The angular momentum of the complete molecule is the resultant of the angular momentum of the nuclei rotating as a rigid body together with that of the electrons around the nuclei, the resultant being formed according to the quantum mechanical rules for the composition of angular momenta. The electronic angular momentum comprises orbital and spin parts. In very many molecules the electrons are paired so that both orbital and spin angular momenta of the electrons are zero and then the angular momentum of the molecules is equal to that of the nuclei. Of the molecules in which electronic angular momentum is not zero, OH and CH are important examples.

Rotating molecules have energy of rotation. If the electrons have no angular momentum, the rotational energy is just that of the nuclei. The simplest case is that of diatomic molecules which have just one moment of

2.1 General remarks

inertia, that about any axis perpendicular to the join of the nuclei; the moment about the inter-nuclear axis is zero. Let the moment of inertia be I. The angular momentum is $I\omega$ where ω is the angular velocity about any axis perpendicular to the join of the nuclei. The quantum condition which gives the possible values of ω is

$$I\omega = N\hbar,$$

where N is an integer.

The rotational energy is $\frac{1}{2}I\omega^2$, so that the possible values are proportional to the square of the angular momentum, that is, to $N(N+1)$ and are

$$\frac{\frac{1}{2}N(N+1)\hbar^2}{I}.$$

If the molecule has an electric dipole moment, a change of angular velocity corresponds to absorption or radiation of electromagnetic energy in which (for dipole radiation) N changes to $N-1$ (emission) or $N+1$ (absorption). Those changes generate the pure rotational spectrum of a diatomic molecule. Since the energies are proportional to I^{-1}, the molecules with the larger moments of inertia have the lower energies and the lower frequencies of transition. The frequencies of hydrides, generally the molecules with the smallest moments of inertia, lie in the far infra-red region of 50 to 500 cm^{-1} ($(1.5-15) \times 10^{12}$ Hz) whereas those of most other molecules lie in the millimetre region of the radio spectrum. The pure rotational spectrum can only occur if the molecule has a high dipole moment; in particular diatomic molecules with the same nuclei (homonuclear) have no dipole moment and show no pure rotational spectrum.

Many molecules of interest to galactic astronomy depart from the simple case just set out, for the system of energy levels is more complex if either the electrons have angular momentum (as with OH) or if the nuclei have more than one moment of inertia (as with H_2O). The complications that then arise will be described below (section 2.2).

So far it has been tacitly assumed that the nuclei are rigidly fixed in their relative positions, but that is not so. If subject to a change in force they will change their positions. First, as they rotate, they are subject to inertial forces, which lead to the nuclei being further apart, and hence to the moment of inertia being greater, as the angular velocity increases. Thus the rotational energy is not simply

$$\frac{\frac{1}{2}N(N+1)\hbar^2}{I}$$

because I varies with N. The effect is rather small but does affect the frequencies of radio transitions.

Secondly, the forces on the nuclei depend on the electronic state, and so I and hence the energies, depend on the electronic state. No galactic radio frequency radiation comes from an electronically excited state of any molecule.

The third consequence of finite forces is that nuclei may vibrate about their mean position. In diatomic molecules the mean separation varies harmonically with an energy of vibration equal to

$$(v+\tfrac{1}{2})E_0$$

where v is the vibrational quantum number and $\tfrac{1}{2}E_0$ is the energy of the ground state. The vibrational energies of most molecules correspond to frequencies in the near infra-red, visible or ultra-violet (again, the lighter the molecule the higher the frequency in general). Now the proportion of molecules in the state of quantum number v is

$$\exp[-(v+\tfrac{1}{2})E_0/kT]$$

and because E_0 is relatively high, corresponding to a radiation temperature of some 10^4 K, the number of molecules in even the first excited state ($v = 1$) will be very low in a galactic gas at a temperature of about 100 K.

Molecules in the galaxy are thus not normally in vibrationally or electronically excited states and radio frequency transitions between rotational levels of electronically excited molecules have never been observed. However molecules may briefly be excited into higher electronic or vibrational states by radiation or collisions and such excitations may play a part in leading to inversions of populations between lower lying levels.

In general the energy of a molecule depends on its electronic state, the vibration of its nuclei and the angular momentum of nuclei and electrons; hyperfine interactions between nuclei and electrons may also occur. In principle none of these phenomena is independent of the others but in many instances it is possible to consider them separately to a good first approximation. The basis of that possibility is the Born–Oppenheimer approximation which depends on the idea that the velocities of the electrons are very much greater than those of the nuclei so that to a first approximation the latter may be ignored in discussing the former. Then to the first approximation the vibratory and rotational motions are independent of the states of the electrons of the molecule and so the wave

2.2 The structure and spectrum of hydroxyl

function of a particular state of the molecule may to the same approximation be written as the product of electronic, vibrational and rotational parts

$$\psi = \psi_{el}\frac{1}{r}\psi_{vib}\psi_{rot}.$$

Correspondingly, the total energy is the sum of three terms:

$$E = E_{el} + E_{vib} + E_{rot}.$$

These decompositions are justified by the theory given in appendix 1.

The vibratory motion of the nuclei of diatomic molecules is along the inter-nuclear axis, while in polyatomic molecules it may be resolved into normal modes.

Diatomic molecules have a negligible moment of inertia of the nuclei about the inter-nuclear axis and a finite one about all (equivalent) axes of inertia perpendicular to the inter-nuclear axis and the angular momentum of the nuclei is equal to $N\hbar$, where N may be zero or any positive integer.

In general, a diatomic molecule will have also a component of electronic orbital angular momentum about the inter-nuclear axis equal to $\Lambda\hbar$, where Λ is an integer, and electronic spin angular momentum, $\Sigma\hbar$, where Σ is an integer or half an integer. In fact, the electrons in the ground electronic states of most molecules are paired so that Λ and Σ are zero; if Λ is zero, the molecule is said to be in a Σ state; if Λ is 1, the molecule is in a Π state (corresponding to the S, P ... notation for the electronic orbital angular momentum of atoms).† The hydroxyl molecule, OH, and the CH molecule are important exceptions to the general rule that the ground states of molecules are Σ states: the ground states are Π states with one unpaired electron and $\Lambda = 1$ and $\Sigma = \frac{1}{2}$.

2.2. The structure and spectrum of hydroxyl

Because hydroxyl is of diatomic molecules the one that emits much the strongest stimulated emission, the further discussion of diatomic molecules is confined to the properties of hydroxyl (those of CH are similar).

The total orbital angular momentum, K, of nuclei and electrons can take any of the values $(\Lambda + N)\hbar$, where Λ is 1 in the ground state and N is 0 or any positive integer. The least value is therefore $\hbar$ (see figure 2.1). The total angular momentum, orbital plus spin, is $J\hbar$, where J can take the

† Note that Σ is used to denote both the resultant spin quantum number and the value of the component Λ of resultant angular momentum.

Molecular structure and spectra

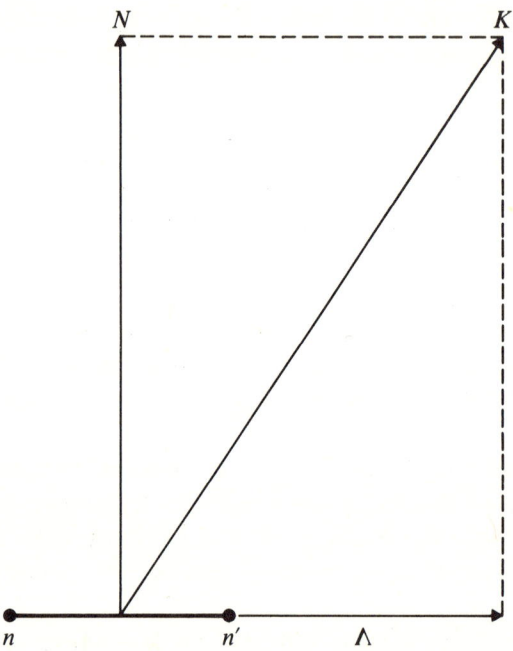

Figure 2.1. Vector diagram of nuclear and electronic orbital angular momentum of diatomic molecules.

values $\Lambda + N \pm \Sigma$. If the negative sign is taken, the values of J form the sequence

$$\tfrac{1}{2}, \tfrac{3}{2}, \tfrac{5}{2} \ldots.$$

Of the two sets of states in hydroxyl, those with the positive sign for Σ, that is those with the sequence

$$J = \tfrac{3}{2}, \tfrac{5}{2}, \tfrac{7}{2} \ldots,$$

have the lower energies (figure 2.2).

The rotational states of the lowest electronic state are labelled $^2\Pi$, the superscript 2 indicated that there are two sequences. The sequence is shown by a suffix equal to the least value of J:

$$^2\Pi_{3/2} \quad \text{or} \quad ^2\Pi_{1/2},$$

and if need be, the actual value of J is added.

$$^2\Pi_{\tfrac{3}{2}}, J = \tfrac{5}{2} \quad \text{or} \quad ^2\Pi_{\tfrac{1}{2}}, J = \tfrac{3}{2}.$$

The foregoing summary of molecular structure depends on the Born–Oppenheimer approximation (Herzberg, 1950; van Vleck, 1929) that the

2.2 The structure and spectrum of hydroxyl

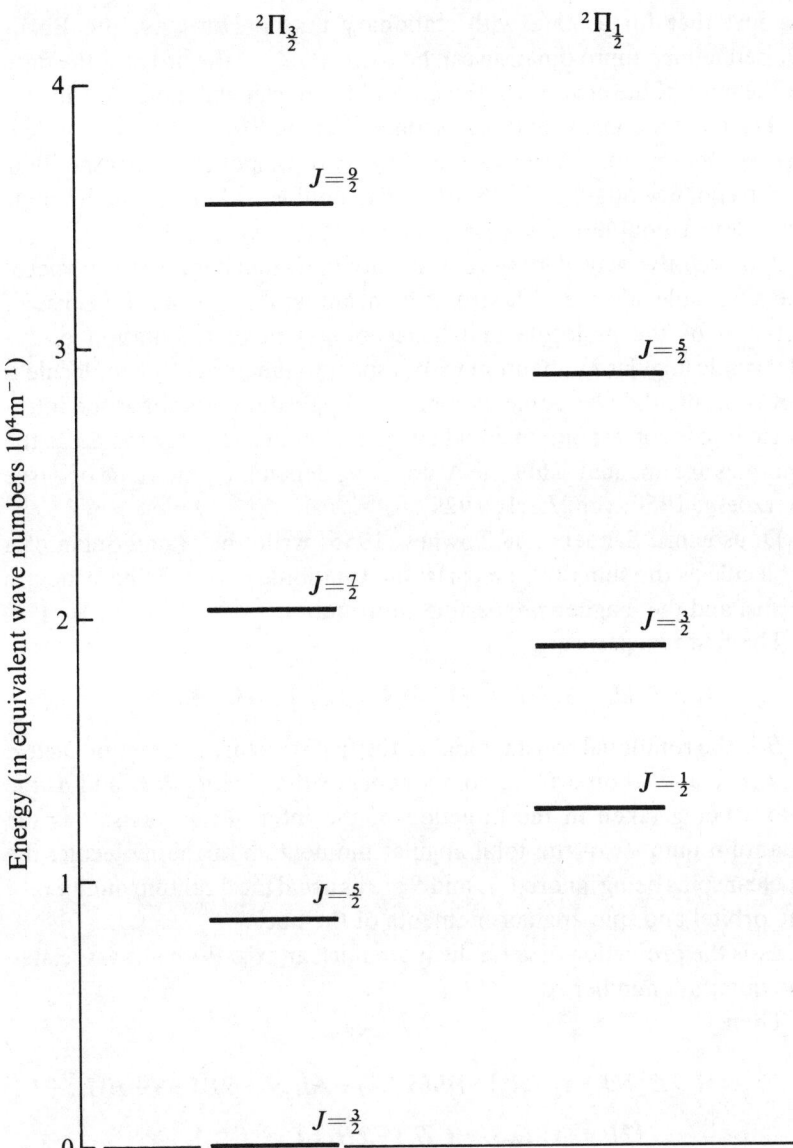

Figure 2.2. Energy levels of the two series of OH states.

velocities of the nuclei are so small relative to those of the electrons, that the wave functions for the electrons may be calculated as if the nuclei were stationary. It is on that basis that the total wave function may be factorized into electronic, vibrational and rotational parts and were it strictly true, the electronic wave function for an actual ground state would be just that for a state with stationary nuclei. However, the Born–Oppenheimer approximation can be correct only to the order of the ratio of the mass of the electron to the mass of the nuclei, that is about 1 in 10^4, and in a better approximation it is found that the $^2\Pi_{\frac{3}{2}}$, $J=\frac{3}{2}$ ground state is actually a doublet, the separation of the two components corresponding to a frequency of about 1666 MHz. The doublet is known as a Lambda Doublet (Λ-doublet).

A descriptive way of looking at Λ-doubling is that because the nuclei of the OH molecule are different, it is meaningful to speak of a sense of rotation of the molecule and hence of a sign of orientation of the electronic angular momentum with respect to the nuclei. If a molecule is not rotating, the electronic momenta of opposite sense about the inter-nuclear axis correspond to equal energies, but in a rotating molecule the energies are unequal. Thus the Λ-doubling depends on the value of J (see Herzberg, 1950; van Vleck, 1929).

Dousmanis, Sanders and Townes (1955) write the Hamiltonian of a molecule as the sum of three parts; the rotational part, the fine structure terms, and the magnetic hyperfine contribution.

The first two parts are:

$$H = B[(J_x - S_x - L_x)^2 + (J_y - S_y - L_y)^2] + A\boldsymbol{L}\cdot\boldsymbol{S}.$$

B is the rotational constant and A the fine structure interaction factor. J_x, L_x, S_x and so on are the components of the vectors $\boldsymbol{J}, \boldsymbol{L}$ and $\boldsymbol{S}$, the z-axis being taken in the direction of the inter-nuclear axis. J is the quantum number of the total angular momentum of the molecule, the nuclear spins being ignored; L and S are as usual the quantum numbers of the orbital and spin angular momenta of the nuclei.

L_z is the projection of $\boldsymbol{L}$ on the inter-nuclear axis. With it is associated the quantum number Λ.

Then

$$H = B[J(J+1)-\Lambda^2] + BS(S+1) + AL_zS_z - 2B\boldsymbol{J}\cdot\boldsymbol{S} + B(L_x^2+L_y^2)$$
$$+ (2B+A)(L_xS_x + L_yS_y) - 2B(J_xL_x + J_yL_y).$$

If the Born–Oppenheimer approximation were to hold, the Hamiltonian would depend only on J, Λ and S, that is it would include only the first four terms. The term $B(L_x^2+L_y^2)$ is effectively a constant, but the last two terms, which depend on the sign of L_x and L_y, generate the Λ-doublet splitting.

The first four terms are diagonal in a representation which includes only the wave functions of the $^2\Pi_{\frac{1}{2}}$ and $^2\Pi_{\frac{3}{2}}$ ground states (appendix 1).

2.2 The structure and spectrum of hydroxyl

Dousmanis, Saunders and Townes (1955, see also, van Vleck, 1929) show how to represent the complete Hamiltonian in diagonal form on a basis which includes the wave functions of the first electronic excited state $^2\Sigma_{\frac{1}{2}}$.

The two levels of a Λ-doublet have opposite parity; in the $^2\Pi_{\frac{3}{2}}$ ground state the upper level has $+$ parity and the lower $-$, in the next level, $J = \frac{5}{2}$, the upper level has $-$ parity, and so on alternating up the $^2\Pi_{\frac{3}{2}}$ sequence; similarly, in the $^2\Pi_{\frac{1}{2}}$ sequence, the upper member of the $J = \frac{1}{2}$ doublet has $-$ parity, that of the $J = \frac{3}{2}$ doublet has $+$ parity, and so on.

There is a further contribution to the energy of the hydroxyl molecule, the hyperfine interaction between the electrons and the proton which depends on the total quantum number F, equal to the vector sum of J and I, where I is the spin quantum number of the proton, namely $\frac{1}{2}$. The two possible values of F are thus

$$J + \tfrac{1}{2} \quad \text{and} \quad J - \tfrac{1}{2}.$$

In the ground state $J = \frac{3}{2}$, and F can be 2 or 1. The levels with $F = 2$ have the higher energy.

The ground state $^2\Pi_{\frac{3}{2}}$, $J = \frac{3}{2}$ is accordingly split by the combination of Λ-doubling and hyperfine interaction into a quartet of levels as follows

Level	Parity	F	Energy (in MHz)
4	$+$	2	1720
3	$+$	1	1665
2	$-$	2	53
1	$-$	1	0

All transitions from upper (4, 3) to lower (2, 1) levels are allowed electric dipole transitions except in the $^2\Pi_{\frac{1}{2}}$, $J = \frac{1}{2}$ state, for which the values of F are 0 and 1. Transitions with $F = 0$ to $F = 0$ are not allowed, and so the transitions in that state are the three $4 \rightarrow 2$, $4 \rightarrow 1$, $3 \rightarrow 2$.

The quartet of levels in the lower rotational states of hydroxyl are listed in table 2.1. The structure of CH is similar.

The frequencies of most of the transitions in table 2.1 have been measured to high precision (Radford, 1961, 1962).

In a small magnetic field B the energy of a level with total quantum number F becomes

$$E_0 + B\mu_B g_F m_F$$

Molecular structure and spectra

Table 2.1. *Details of some of the lowest energy levels in the $^2\Pi_{\frac{3}{2}}$ and $^2\Pi_{\frac{1}{2}}$ rotational ladders of hydroxyl (Burdyuzha and Varshalovich, 1973)*

State	Energy above ground state (m^{-1})	Transitions in quartet ($F_{initial} \to F_{final}$)	Frequency of transition (MHz)	Equilibrium relative line strength in quartet (triplet for $^2\Pi_{\frac{1}{2}}, J=\frac{1}{2}$)
$^2\Pi_{\frac{3}{2}}, J=\frac{3}{2}$	0	$1 \to 2$	1612	1
		$1 \to 1$	1665	5
		$2 \to 2$	1667	9
		$2 \to 1$	1720	1
$^2\Pi_{\frac{3}{2}}, J=\frac{5}{2}$	8400	$2 \to 3$	6017	1
		$2 \to 2$	6031	14
		$3 \to 3$	6035	20
		$3 \to 2$	6049	1
$^2\Pi_{\frac{1}{2}}, J=\frac{1}{2}$	12 600	$0 \to 1$	4660	1
		$1 \to 1$	4751	2
		$1 \to 0$	4766	1
$^2\Pi_{\frac{1}{2}}, J=\frac{3}{2}$	18 800	$1 \to 2$	7749	1
		$1 \to 1$	7761	5
		$2 \to 2$	7820	9
		$2 \to 1$	7832	1
$^2\Pi_{\frac{3}{2}}, J=\frac{7}{2}$	20 200	$3 \to 4$	13 434	1
		$3 \to 3$	13 435	27
		$4 \to 4$	13 441	35
		$4 \to 3$	13 442	1
$^2\Pi_{\frac{1}{2}}, J=\frac{5}{2}$	28 900	$3 \to 2$	8118	1
		$2 \to 2$	8136	14
		$3 \to 3$	8190	20
		$2 \to 3$	8207	1
$^2\Pi_{\frac{3}{2}}, J=\frac{9}{2}$	35 500	$4 \to 5$	23 805	1
		$4 \to 4$	23 818	44
		$5 \to 5$	23 827	54
		$5 \to 4$	23 839	1

where

E_0 is the energy in zero field;

μ_B is the Bohr magneton, $e\hbar/2mc$ (e is the charge and m the mass of the electron).

The numerical value of μ_B is 9.2732×10^{-24} J T^{-1} and of μ_B/h, 13.997 GHz T^{-1}.

2.2 The structure and spectrum of hydroxyl

m_F, the magnetic quantum number, takes the values

$$-F, -(F-1)\ldots(F-1), F.$$

g_F is the Lande splitting factor, and is equal to

$$g_J \cdot \frac{F(F+1)+J(J+1)-I(I+1)}{2F(F+1)}.$$

Because m_F can take any one of $(2F+1)$ values, a level with a single energy in zero field splits into $2F+1$ magnetic sub-levels in a small field. Thus if $F=2$ there are five sub-levels and three if $F=1$.

The factor g_J has been measured by Radford (1961). The values, slightly different for the upper and lower levels of the Λ-doublet of the $^2\Pi_{\frac{3}{2}} J = \frac{3}{2}$ state are

upper levels (+parity): 0.935
lower levels (−parity): 0.936

In the upper state ($F=2$) the factor

$$\frac{F(F+1)+J(J+1)-I(I+1)}{2F(F+1)} \text{ is } \frac{3}{4},$$

and in the lower state ($F=1$), it is $\frac{5}{4}$.

The changes of frequency corresponding to the changes of energy of the four Λ-doublet levels of the ground state are thus those shown in table 2.2(a).

The selection rule for electric dipole radiation is that

$$\Delta m_F = 0 \quad \text{or} \quad \pm 1.$$

If the magnetic field is directed along the line of sight, the transitions with $\Delta m_F = \pm 1$ are those seen as radiation with left or right circular polarization. The change of frequency of a transition in a magnetic field is

$$B[(\mu_B g_F m_F)_1 - (\mu_B g_F m_F)_2]$$

with $(m_F)_1 - (m_F)_2$ equal to ± 1.

The values of the constant factor, denoted by γ are given for the ground state transitions in table 2.2(b) and a diagram of the components of the $4 \to 2$ transition of the ground state is given in figure 2.3.

The intensities of spontaneous transitions are calculated as follows. The number of photons emitted per second is proportional to the number of molecules in the upper state multiplied by the probability A_{ij} of a spontaneous transition.

Table 2.2.

(a) Zeeman shifts in the levels of the ground state Λ-doublet of OH

Level	F	Energy (MHz)	m_F	$\mu_B g_F m_F$ (GHz T^{-1})
4	2	1720	0	0
			±1	±9.814
			±2	±19.628
3	1	1665	0	0
			±1	±16.357
2	2	53	0	0
			±1	± 9.828
			±2	±19.656
1	1	0	0	0
			±1	±16.379

(b) Zeeman shifts in the transitions of the ground state Λ-doublet of OH

Transition	Frequency (MHz)	Δm_F	Change of frequency for unit magnetic induction (GHz T^{-1})			
4 → 1	1720	+1	− 3.249	−9.814	−16.379	
		−1	16.379	+9.814	+ 3.249	
4 → 2	1667	+1	−9.800	−9.814	−9.828	−9.841
		−1	+9.841	+9.828	+9.814	+9.800
3 → 1	1665	+1	−16.379	−16.357		
		−1	+16.357	+16.379		
3 → 2	1612	+1	− 3.298	−9.828	−16.357	
		−1	−16.357	+9.828	+ 3.298	

The number of molecules is proportional to the number of magnetic sub-levels, namely $2F+1$, so that the intensity is proportional to

$$(2F+1)A_{ij}.$$

The coefficients, A_{ij}, are related to the matrix elements of the dipole moment of the molecule:

$$A_{ij} = \frac{64\pi^4 \nu_{ij}^3}{3hc^3}|R_{ij}|^2$$

where

$$R_{ij} = \int \psi_i^* M \psi_j \, d\tau$$

and M is the dipole moment, ψ_i and ψ_j the wave functions of the two

2.2 The structure and spectrum of hydroxyl

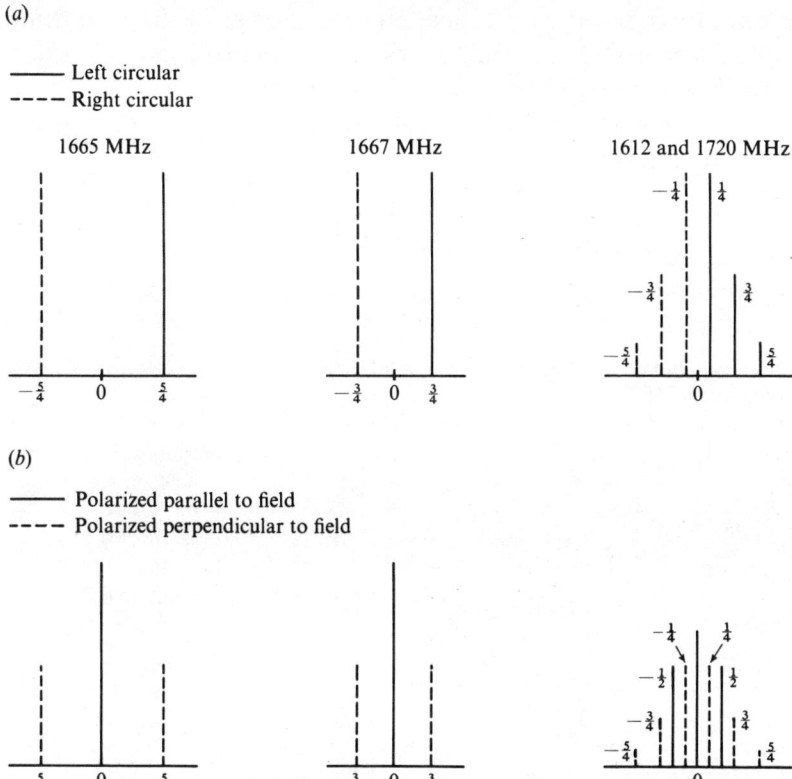

Figure 2.3. Zeeman components of the transitions in the Λ-doublet of the $^2\Pi_{\frac{3}{2}}$, $J = \frac{3}{2}$ state of OH; (a) observation parallel to field, (b) observation perpendicular to field.

states between which the transition occurs, and $d\tau$ is an element of volume.

Similarly, the coefficient of stimulated emission is given by

$$B_{ji} = \frac{8\pi^3}{3h^2c^4}|\boldsymbol{R}_{ij}|^2.$$

The wave functions to be used in calculating $\boldsymbol{R}_{ij}$ are the functions of the upper and lower Λ-doublet states. Burdyuzha and Varshalovich (1973) have made extensive calculations of the transition probabilities for the OH and CH molecules, both for transitions between different rotational levels as well as for transitions between Λ-doublet levels.

In all such calculations, the matrix elements of the dipole moment can be found if the forms of the initial and final wave functions are known – Burdyuzha and Varshalovich (1973) take them to be superpositions of the wave functions of the $^2\Pi_{\frac{1}{2}}$ and $^2\Pi_{\frac{3}{2}}$ states (see appendix 1), it being permissible to neglect the small contribution from the $^2\Sigma_{\frac{1}{2}}$ wave function in calculating matrix elements for transition probabilities. The dipole moment cannot at present be calculated with adequate accuracy; the measured values are

for OH: 1.660 debye (Powell and Lide, 1965),
CH: 1.46 debye (Phelps and Dalby, 1966).

Some of the transition probabilities are given in table 2.3.

Table 2.3. *Transition probabilities in Λ-doublets of OH and CH (Burdyuzha and Varshalovich, 1973)*

Rotational state	Transition $F_1 \to F_2$	Spontaneous probability (s^{-1}) OH	CH
$^2\Pi_{\frac{3}{2}}, J=\frac{3}{2}$	$1 \to 2$	1.29×10^{-11}	4.46×10^{-10}
	$1 \to 1$	7.10×10^{-11}	2.26×10^{-9}
	$2 \to 2$	7.70×10^{-11}	2.46×10^{-9}
	$2 \to 1$	0.94×10^{-11}	2.77×10^{-10}
$^2\Pi_{\frac{1}{2}}, J=\frac{1}{2}$	$0 \to 1$	1.08×10^{-9}	3.07×10^{-10}
	$1 \to 1$	7.63×10^{-10}	2.01×10^{-10}
	$1 \to 0$	3.85×10^{-10}	1.07×10^{-10}
$^2\Pi_{\frac{3}{2}}, J=\frac{5}{2}$	$2 \to 3$	1.09×10^{-10}	
	$2 \to 2$	1.53×10^{-9}	
	$3 \to 3$	1.57×10^{-9}	
	$3 \to 2$	7.89×10^{-11}	
$^2\Pi_{\frac{1}{2}}, J=\frac{3}{2}$	$1 \to 2$	1.86×10^{-10}	5.65×10^{-14}
	$1 \to 1$	9.36×10^{-10}	3.13×10^{-13}
	$2 \to 2$	1.03×10^{-9}	3.95×10^{-13}
	$2 \to 1$	1.15×10^{-9}	4.82×10^{-14}
$^2\Pi_{\frac{3}{2}}, J=\frac{7}{2}$	$3 \to 4$	3.39×10^{-10}	
	$3 \to 3$	9.16×10^{-9}	
	$4 \to 4$	9.25×10^{-9}	
	$4 \to 3$	2.64×10^{-10}	

2.3. Polyatomic molecules

Of the three principal moments of inertia which generally characterize a body, the one along the inter-nuclear axis is zero in diatomic molecules, but the moments of polyatomic molecules are in general all non-zero. Let the principal moments of inertia be denoted by A, B, C. Then if a molecule is symmetrical about some axis, the moment of inertia about

2.3 Polyatomic molecules

that axis will be one of the principal moments and the other two will be equal.

Let C be the moment of inertia about the axis of symmetry and A that about any perpendicular (equatorial) axis. The angular momentum of the molecule has components $C\omega_C$ and $A\omega_A$ where ω_C and ω_A are the angular velocities about the polar and equatorial axes. The quantum condition is that the sum of squares of angular momenta is equal to $J(J+1)\hbar^2$:

$$C^2\omega_C^2 + A^2\omega_A^2 = J(J+1)\hbar^2,$$

or

$$M_C^2 + M_A^2 = J(J+1)\hbar^2.$$

The energy of rotation, E, is

$$\tfrac{1}{2}(C\omega_C^2 + A\omega_A^2),$$

that is

$$\frac{1}{2}\left(\frac{M_C^2}{C} + \frac{M_A^2}{A}\right).$$

Since $M_C^2 = J(J+1)\hbar^2 - M_A^2,$

$$E = \tfrac{1}{2}J(J+1)\hbar^2 + \tfrac{1}{2}M_A^2\left(\frac{1}{A} - \frac{1}{C}\right).$$

The energy is therefore determined by the moments of inertia A and C, the total angular momentum J and the projection of J on the A-axis (or the C-axis). The latter is denoted by K, and can take the values $0 \ldots J$.

Thus for a given value of J, a symmetrical molecule may have one of $J+1$ energies; the levels are labelled J_K.

Water is not axisymmetric but triaxial, with three distinct moments of inertia.

Triaxial molecules are considerably more complex. Let the three moments be A, B and C and the angular velocities about them be ω_A, ω_B, ω_C.

Then the energy, E, is

$$\tfrac{1}{2}(A\omega_A^2 + B\omega_B^2 + C\omega_C^2) = \tfrac{1}{2}\left(\frac{M_A^2}{A} + \frac{M_B^2}{B} + \frac{M_C^2}{C}\right).$$

The quantum condition is

$$M_A^2 + M_B^2 + M_C^2 = J(J+1)\hbar^2,$$

Molecular structure and spectra

so that

$$E = \frac{\hbar}{2}\left[\frac{J(J+1)}{C} + \Lambda_1^2\left(\frac{1}{A} - \frac{1}{C}\right) + \Lambda_2^2\left(\frac{1}{B} - \frac{1}{C}\right)\right].$$

Three parameters are now required to specify the rotational state. The usual scheme is to consider how the states of either a prolate spheroidal molecule $(A > C)$ or an oblate spheroidal molecule $(C > A)$ would be split if the axes A and B were not exactly equal, when it is possible to label the levels of a triaxial molecule by the values of K for the corresponding prolate and oblate axisymmetrical molecules namely K_A and K_C (figure 2.4). The specification of the rotational state of a molecule then reads

$$J_{K_A K_C}.$$

For example, the upper and lower states of the 1.35 cm transition in water are respectively 6_{16} and 5_{23}.

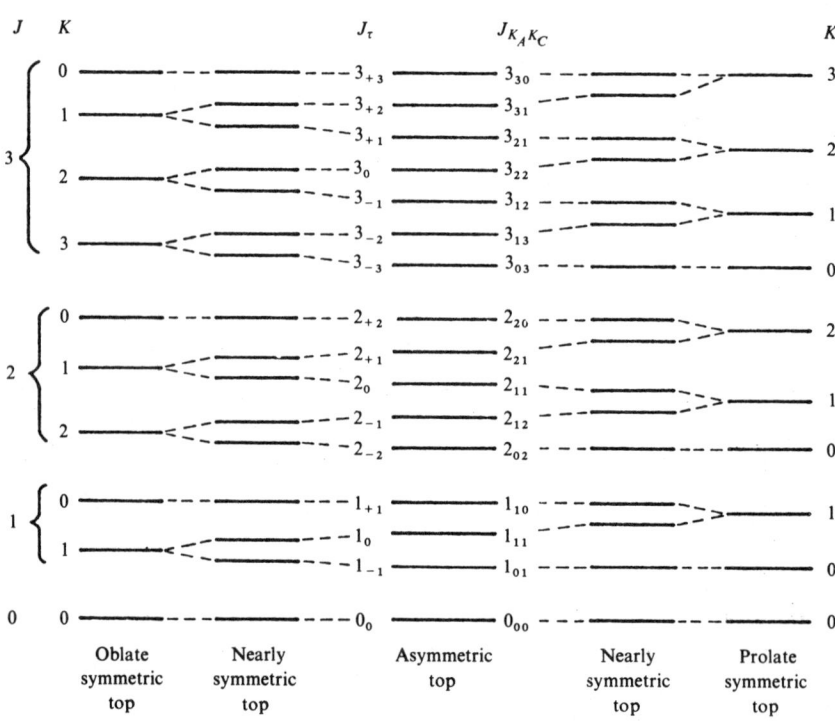

Figure 2.4. Correspondence of the labels K_A and K_C of a triaxial molecule to the labels K in prolate and oblate molecules. (ter Haar and Pelling. © 1974. The Institute of Physics.)

2.4 Intensities and selection rules

Water is a top with rotational constants

$$A = 2788 \text{ m}^{-1}, \quad B = 1451 \text{ m}^{-1}, \quad C = 929 \text{ m}^{-1},$$

and is thus highly asymmetric.

Energies of some of the pure rotational levels of water are listed in table 2.4 and a diagram of some of the levels is shown in figure 2.5.

Table 2.4. *Energies of some rotational levels in water*

Level	Energy (m^{-1})
1_{01}	2379
1_{10}	4236
2_{12}	7950
2_{21}	13 450
3_{03}	13 676
3_{12}	17 337
3_{21}	21 216
3_{30}	28 542
4_{14}	22 484
4_{23}	30 037
4_{32}	38 252
4_{41}	48 811
5_{05}	32 540
5_{14}	39 951
5_{23}	44 656
5_{32}	50 886
5_{41}	61 038
5_{50}	74 212
6_{16}	44 730
6_{25}	55 296
6_{34}	64 903
6_{43}	75 676

2.4. Intensities and selection rules for rotational transitions

As before, the Einstein coefficient for transitions between two rotational states i and j is

$$A_{ij} = \frac{64\pi^4 \nu_{ij}^3}{3hc^3} |\mathbf{R}_{ij}|^2$$

Molecular structure and spectra

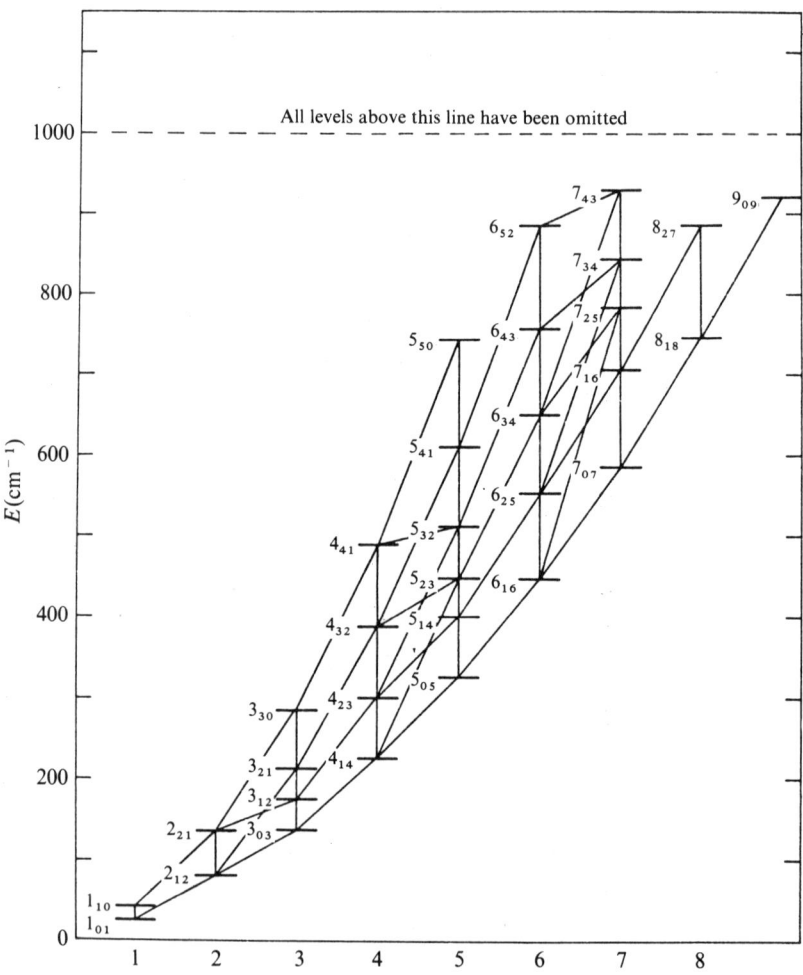

Figure 2.5. Partial term diagram of rotational states of water. Abscissa: values of J. Ordinate: energy in cm^{-1}.

where

$$|R_{ij}| = \int \Psi_i M \Psi_j \, d\tau$$

and Ψ_i and Ψ_j are the wave functions of the respective states.

When Ψ_i and Ψ_j are taken to be the wave functions for the top (appendix 1) the following selection rules follow:

In every case, $\Delta J = 0, \pm 1$ except that $J = 0 \rightarrow J = 0$ is forbidden.

For a symmetric top (axisymmetric molecule)

$$\Delta K = \pm 1.$$

2.5 Hyperfine components in water

For an asymmetric top (triaxial molecule) there are two possibilities, depending on the direction of the dipole moment relative to the principal axes.

Either $\Delta K_A = 0, \pm 2$ and $\Delta K_C = \pm 1, \pm 3$ as for formaldehyde *or* $\Delta K_A = \pm 1, \pm 3$, and $\Delta K_C = \pm 1, \pm 3$ as for water.

In the 1.35 cm line of water, $\Delta J = -1$, $\Delta K_A = +1$, $\Delta K_C = -3$.

The dipole moment of water (directed along the intermediate axis) is 1.88 debye, a high value, and the intensities of rotational transitions are high. Some values are given in table 2.5. It should be particularly noted that while many transitions have probabilities between 10^{-3} and 10^{-1} s^{-1}, the one which shows maser action, 6_{16}–5_{23}, has a probability only about 100 times greater than the probabilities of the Λ-doublet transitions in OH; the reason is that the frequency enters the transition probability as the cube, and the frequency of the 6_{16}–5_{23} transition is much less than the others listed in table 2.5.

Table 2.5. *Probabilities of some rotational transitions in water*

Transition	Probability (s^{-1})
1_{01}–1_{10}	3.55×10^{-3}
2_{21}–1_{10}	2.60×10^{-1}
2_{21}–2_{12}	3.08×10^{-2}
3_{30}–2_{21}	1.29
3_{30}–3_{03}	8.60×10^{-3}
4_{23}–4_{14}	8.34×10^{-2}
4_{32}–4_{23}	1.26×10^{-1}
5_{14}–4_{23}	1.61×10^{-1}
5_{24}–5_{05}	7.80×10^{-2}
5_{23}–4_{32}	1.15×10^{-2}
6_{16}–5_{23}	1.91×10^{-9}

2.5. Hyperfine components in water

The quantum number I of the total nuclear spin in water is 1, but the range of the five hyperfine components of the 6_{16}–5_{23} transition is only 200 kHz or 10^{-5} of the frequency (Bluyssen, Dymanus and Verhoeven, 1969) so that the components are not separated in astronomical observations, although there is some evidence that they affect the profiles of the lines that are observed.

3
Observations of celestial masers

3.1. Introduction – the general properties of OH and H_2O masers

When a maser source is observed with a single radio receiver, it is found that the source cannot be resolved, that is, it lies wholly within the beam of a single aerial and the spectra of the four OH lines as determined with such a receiver are found to have typically the character shown in figure 3.1. It was explained in chapter 1 that the intensity of the signal emitted by the maser cannot be found from the intensity of the signal received at a single aerial because the angular size of the source is unknown (it is only known to be less than the solid angle embraced by the aerial beam). It is the intensity of the radiation received by the aerial that is observed and that is usually quoted as the equivalent temperature of the aerial. The aerial temperature can be displayed as a function of frequency, but because it is almost invariably supposed (for want of a convincing alternative) that any displacement of frequency is a Doppler shift consequent on a mass motion of the radiating atoms along the line of sight from the Earth, differences of frequency from the frequency emitted by molecules at rest on the Earth are expressed as equivalent velocities. Since the frequency of hydroxyl radiation is about 1650 MHz, a difference of 10 kHz corresponds to a line of sight velocity of about 1.8 km s^{-1}.

Clouds of water molecules radiate by stimulated emission, but only one transition can be detected at radio frequencies (as pointed out in chapter 2, the hyperfine levels are too close to be distinguished). The spectra are, like those of hydroxyl, made up of a number of components each with a relative width of one part in a million or thereabouts; they do not in general show the high degree of circular polarization (almost 100 per cent) observed in hydroxyl sources, but display partial linear polarization (section 3.6).

A remarkable feature of some water sources is the wide extent of the spread of frequency, corresponding to a spread of mass velocities of some 500 km s^{-1} (figure 3.2).

It was found fairly soon after the first detection of hydroxyl maser sources that they varied in time (Weaver, Dieter and Williams, 1968;

3.1 Introduction

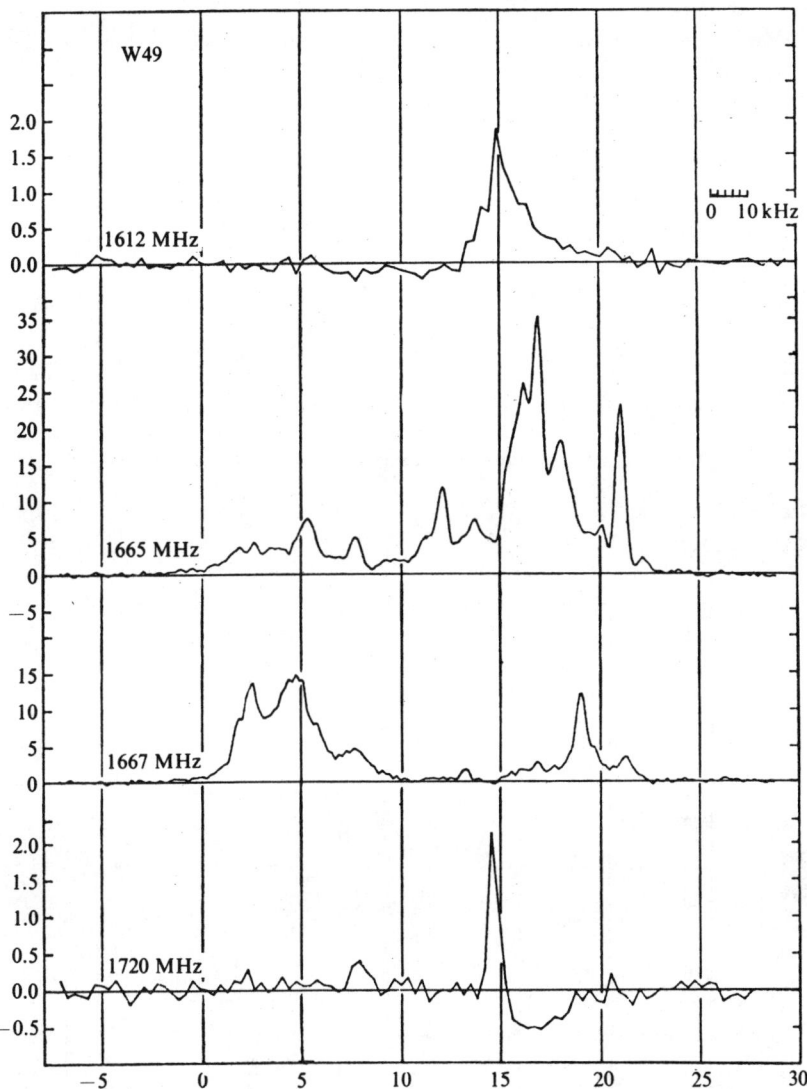

Figure 3.1. Representative set of hydroxyl spectra – W49 (Weaver, Dieter and Williams. © 1968. The University of Chicago). Abscissa: velocity (km s^{-1}) equivalent to Doppler shift of frequency. Ordinate: aerial temperature (K), proportional to intensity.

Zuckerman, Ball, Dickinson and Penfield, 1969). As may be seen in figure 3.3, the general level of intensity of the signal may change and in addition the shapes of the spectra may alter, corresponding to changes in the relative intensities of components and indeed, while some components increase in intensity, others decrease (see also, Wilson, Davies

Observations of celestial masers

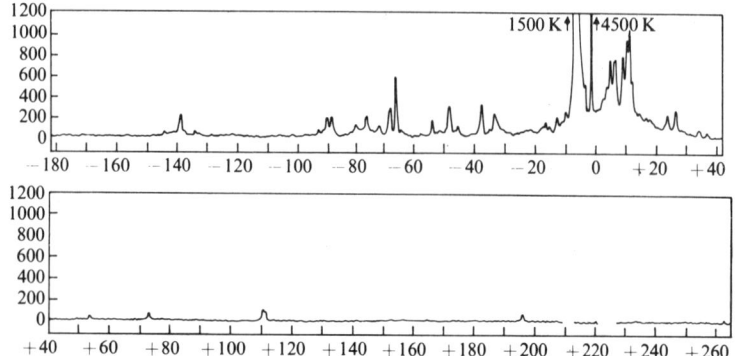

Figure 3.2. Water spectrum of W49 (Sullivan. © 1971. The University of Chicago). Abscissa: velocity (km s^{-1}) equivalent to Doppler shift of frequency. Ordinate: aerial temperature (K), proportional to intensity.

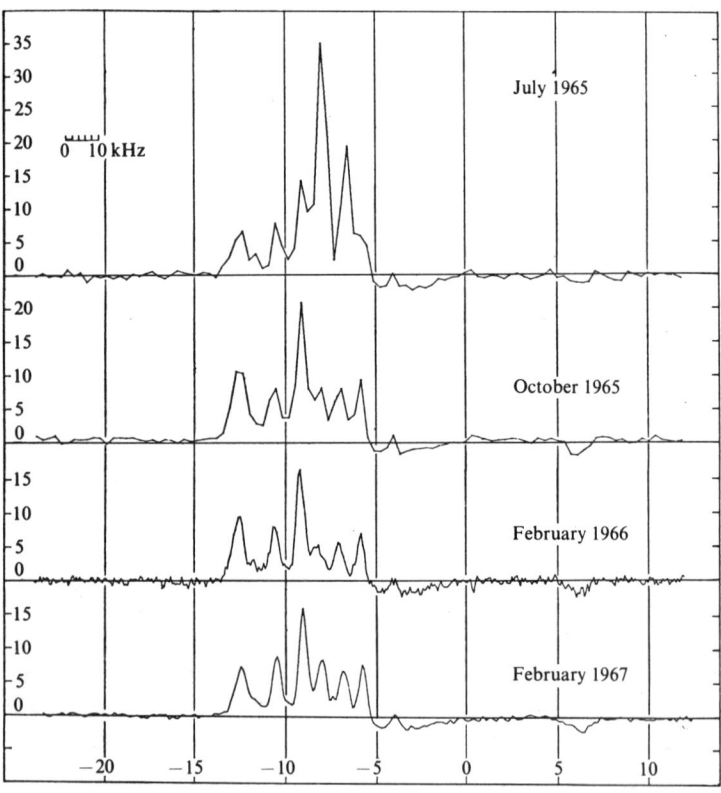

Figure 3.3(a)

3.1 Introduction

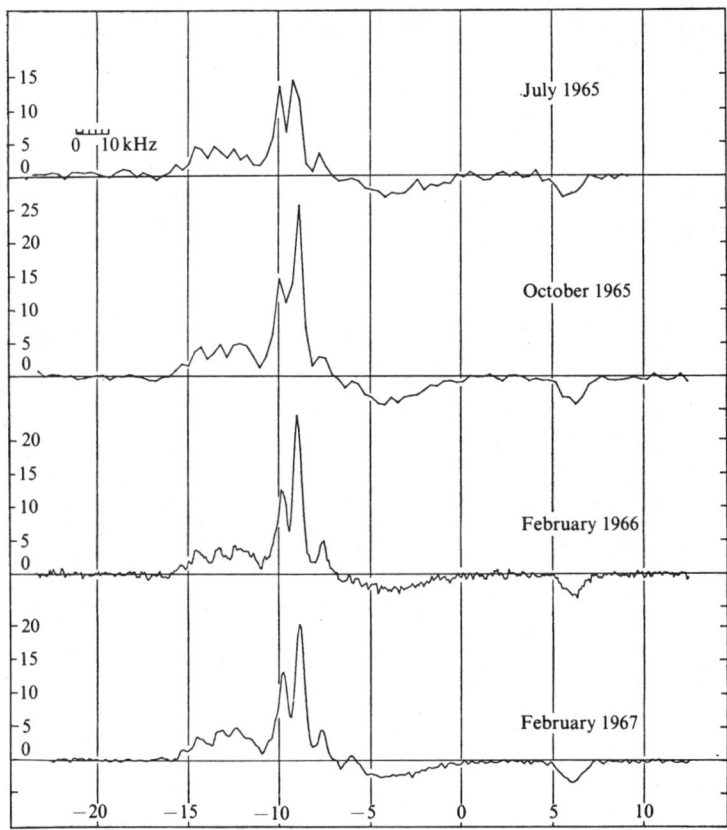

Figure 3.3(b)

Figure 3.3. Variations of spectra of hydroxyl sources in NGC 6334 with time (Weaver, Dieter and Williams. © 1968. The University of Chicago); (a) 1667 MHz, (b) 1665 MHz. Abscissa: velocity (km s^{-1}) equivalent to Doppler shift of frequency. Ordinate: aerial temperature (K), proportional to intensity.

and Ellder, 1972). Intensities of water masers may change even more rapidly than those of hydroxyl masers (figure 3.4; Buhl and others, 1969).

Studies with two-aerial interferometers show that the individual components of the spectra are radiated from very small regions (Moran and others, 1968; Burke and others, 1972; Moran and others, 1973). Those of hydroxyl masers may be 10^{-2} arc sec across or less, and those of water masers as small as 10^{-4} arc sec (see table 3.1).

Let a source have an area A when projected on to the plane normal to the line of sight from a receiving aerial. Let its brightness be B_ν, so that the

31

Observations of celestial masers

total power emitted into a solid angle dω is

$$AB_\nu \, d\omega \, d\nu,$$

dν being the bandwidth of the radiation.

Let the area of the receiving aerial be a and let the distance of the source from the aerial be D. Then the aerial subtends an angle d$\omega = a/D^2$ at the source and the power received by the aerial is

$$AB_\nu a D^{-2} \, d\nu.$$

Suppose the polar diagram of the aerial to have a uniform response over a solid angle Ω. The apparent area of the source, assumed to be

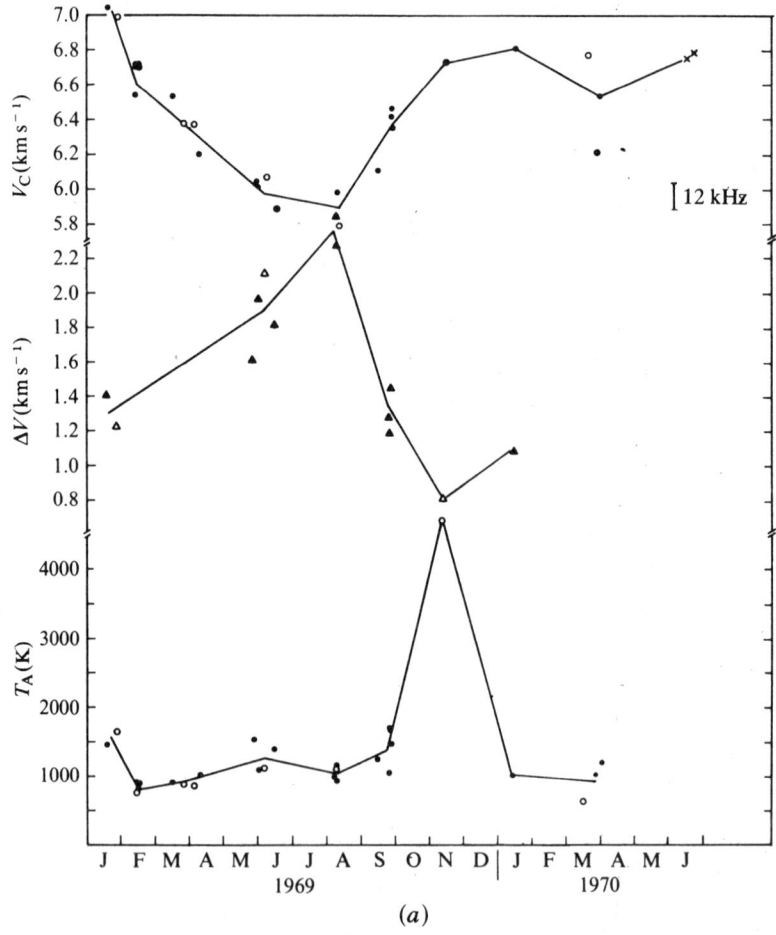

Figure 3.4. Variations of water source with time (Sullivan. © 1971. The University of Chicago): (a) Radial velocity, v, width, Δv, and corrected antenna temperature, T_A, of the $+6.5$ km s^{-1} H$_2$O feature in W49 (in

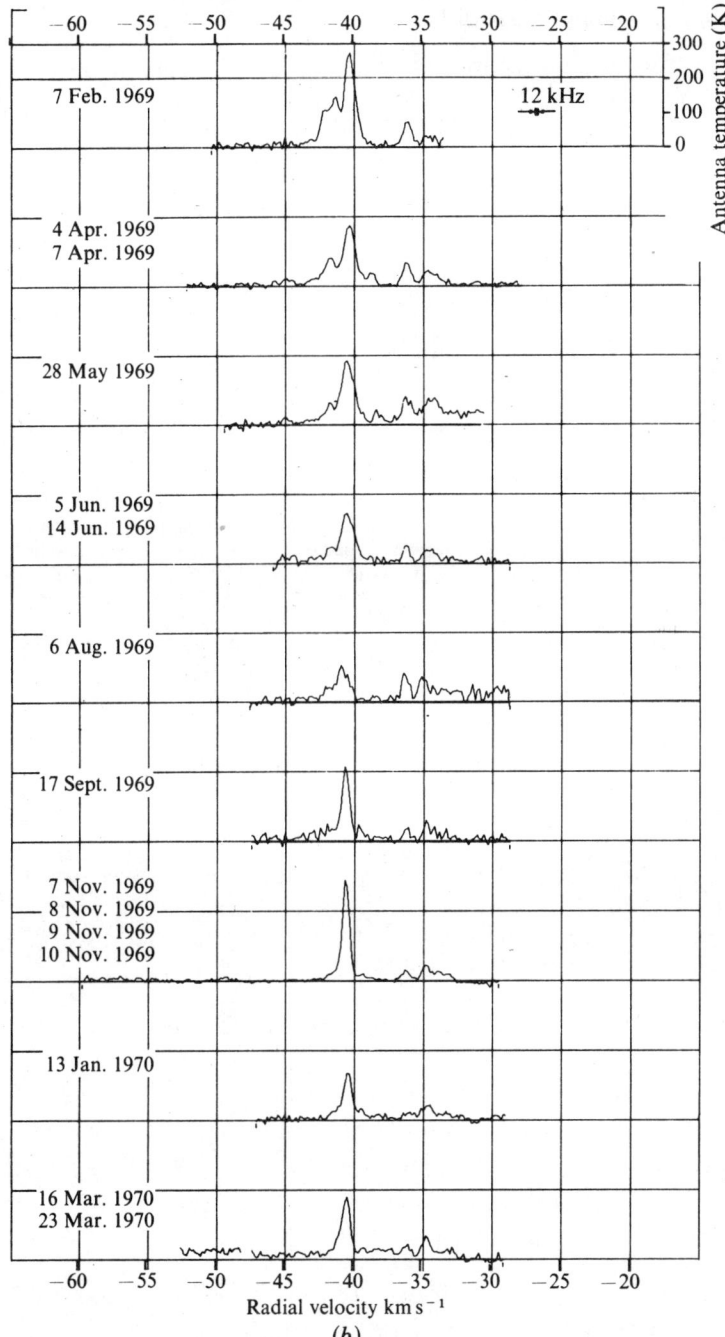

various polarization modes) over the period 1969 January to 1970 June. (b) Examples of H_2O profiles for W3 (feed angle = 0°) over the period 1969 February to 1970 March taken with the Maryland Point 85-foot telescope.

Observations of celestial masers

Table 3.1. *Sizes and brightnesses of masers*

(a) Hydroxyl masers (Harvey and others, 1974)

Source	Distance (kpc)	Size of source region (arcsec)	(10^{11} km)	Diameter of elementary source (10^{-3} arcsec)	(10^8 km)	Brightness temperature (10^{12} K)
W49(1)	14.1†	3.3×1.2	70×25	50	100	
W49(2)	14.1†	1.8×0.8	38×17			
W3(OH)	2.6	2.3×1.4	9×5	5–20‡	20–75‡	2
W75(S)	1.5	0.2×0.2	0.5×0.5			
W75(N)	1.5	1.0×0.2	2.5×0.5			
Sgr B2	10.0	2.4×0.7	36×11			
VY CMa (1667–1665)	1.5†	0.5×0.5	1×1			

(b) Water masers (Moran and others, 1973)

Source	Distance (kpc)	Size of source region (arcsec)	(10^{11} km)	Diameter of elementary source (10^{-3} arcsec)	(10^8 km)	Brightness temperature (10^{12} K)
W49	14.1	1.1×1	20	0.3	6.7	1500
W3(OH)	2.6	1.9×0.3	7	<2	<7.5	>3
Ori A	0.5	13.6×12.1	20	0.8	0.6	35
VY CMa	1.5	about 0.2	0.3	<2	<5	>3

† Moran and others (1973) give distances of 15 kpc and 1 kpc for W49 and VY CMa respectively.

‡ W3(OH) is the only source for which VLBI results are available to give the diameters of individual spots.

unresolved, is then ΩD^2 and the apparent brightness of the source would be calculated to be

$$B_\nu = \frac{A B_\nu a D^{-2}}{d\nu\, a D^{-2} \Omega D^2} = B_\nu \cdot \Omega_s/\Omega$$

where $\Omega_s = A/D^2$ is the true solid angle subtended by the source at the aerial.

Now the aerial temperature T_a is related to the apparent brightness of the source by Planck's formula, the source being assumed to radiate as a black body:

$$B'_\nu = \frac{2h\nu^3}{c^2\{\exp(h\nu/kT_a) - 1\}}.$$

Even if the brightness is such that T_a is 1 K, $h\nu/kT_a$ is about 0.1 when ν is 1660 MHz and so B'_ν may be written as

$$\frac{2\nu^2}{c^2} kT_a \quad \text{or as} \quad \frac{2kT_a}{\lambda^2}.$$

At higher temperature the approximation is better.

3.1 Introduction

It follows that if T_s is the temperature corresponding to the brightness of the source, then

$$T_s = T_a B_\nu / B'_\nu = T_a \Omega / \Omega_s.$$

It must be emphasized that maser sources do *not* radiate as black bodies. The radiation is restricted to a very narrow band of frequencies, of the order of one part in a million of the centre frequency of the line. The temperatures, T_a or T_s, are those which the source or aerial would have to have if either were a black body and if the power radiated at the frequency of the line and within the same spread of frequency, were to be the same as that observed in the actual narrow line. It is not of course necessary to know Ω if the source lies wholly within the response of the polar diagram. The total power P_r received by the aerial was seen to be

$$AB_\nu a D^{-2} \, d\nu$$

and that is the quantity actually measured. In this expression, a is known, A is estimated from interferometric observations, and if D can be estimated for example from the known distance of an associated object, an H-II region or an infra-red object, then

$$B_\nu = \frac{1}{d\nu} \frac{P_r D^2}{Aa}$$

and

$$T_s = \frac{c^2 B_\nu}{2k\nu^2}.$$

A typical OH source may have an equivalent black body temperature of 10^{12} K, even though it radiates over a bandwidth of no more than about 2 kHz. The surface brightness is accordingly about 8×10^{-11} W m^{-2} Hz^{-1} sterad^{-1}, and the power radiated into a bandwidth of 2 kHz about 1.5×10^{-7} W m^{-2} sterad^{-1}.

The separation of the components of a source is simplified if the elements can be resolved in direction by interferometric observations and if components are distinguished by sense of polarization.

Now the natural width of the hydroxyl radiation, for which the probability of transition is 10^{-11} s^{-1}, is about 10^{-11} Hz and negligible compared with the observed width of a few kilohertz. Further, the density of molecules is probably low enough that broadening of the line by collisions is also negligible. The observed width must then correspond to the range of kinetic velocities of the gas and will be of the order of $2\nu_0 \sigma(v)/c$ where

ν_0 is the nominal frequency, $\sigma(v)$ is the standard deviation of the thermal velocities, and c is the speed of light.

$\sigma(v)$ is $(kT_k/m)^{\frac{1}{2}}$ where T_k is the kinetic temperature and m is the mass of the molecule.

Thus

$$T_k = \frac{m}{k}\left(c \cdot \frac{\delta\nu}{2\nu_0}\right)^2.$$

$m = 2.8 \times 10^{-26}$ kg for hydroxyl and if $\delta\nu/\nu_0$ is 10^{-6},

$T_k = 50$ K, approximately (see p. 4).

The gross discrepancy between temperatures corresponding to the apparent width of the line as radiated and to the observed intensity, is apparent. If radiation comes from spontaneous emission, the kinetic and brightness temperatures are usually of the same order.

While individual components of an hydroxyl or water spectrum have relative widths of about 1 in 10^6, the components occur at frequencies which cover a range of some 150 kHz (see figures 3.1 and 3.3), corresponding to a range of mass velocities in the sources of some 30 km s^{-1}, or 100 times the random thermal velocities as estimated from the line widths. Observations with interferometers (figure 3.5) show that source regions contain a number of elements (typically between 10 and 100) each of which subtends between 10^{-2} and 10^{-4} seconds of arc and has a particular Doppler shift (Cooper, Davies and Booth, 1971; Harvey and others, 1974).

The radiation from each element may be polarized. If the region as a whole is examined with a single aerial it is almost always found that each component of the radiation at 1665 and 1667 MHz from hydroxyl is nearly 100 per cent circularly polarized, either right-handed or left-handed (figure 3.6). Interferometric observations show that radiation from each element of the source region is likewise almost completely polarized in one sense or the other (Cooper, Davies and Booth, 1971; Harvey and others, 1974). The 1612 MHz radiation may also be nearly completely circularly polarized. The observations of the 1720 MHz radiation from hydroxyl are not so clear, while radiation from water is usually found to be partially linearly polarized.

The first hydroxyl maser sources that were discovered were close to H-II regions – clouds of hydrogen that are almost completely ionized by the ultra-violet radiation from one or more hot stars of class O or B within them. Subsequently some hydroxyl sources were found close to infra-red stars or nebulae. H-II regions are, with very few exceptions (of which the

3.1 Introduction

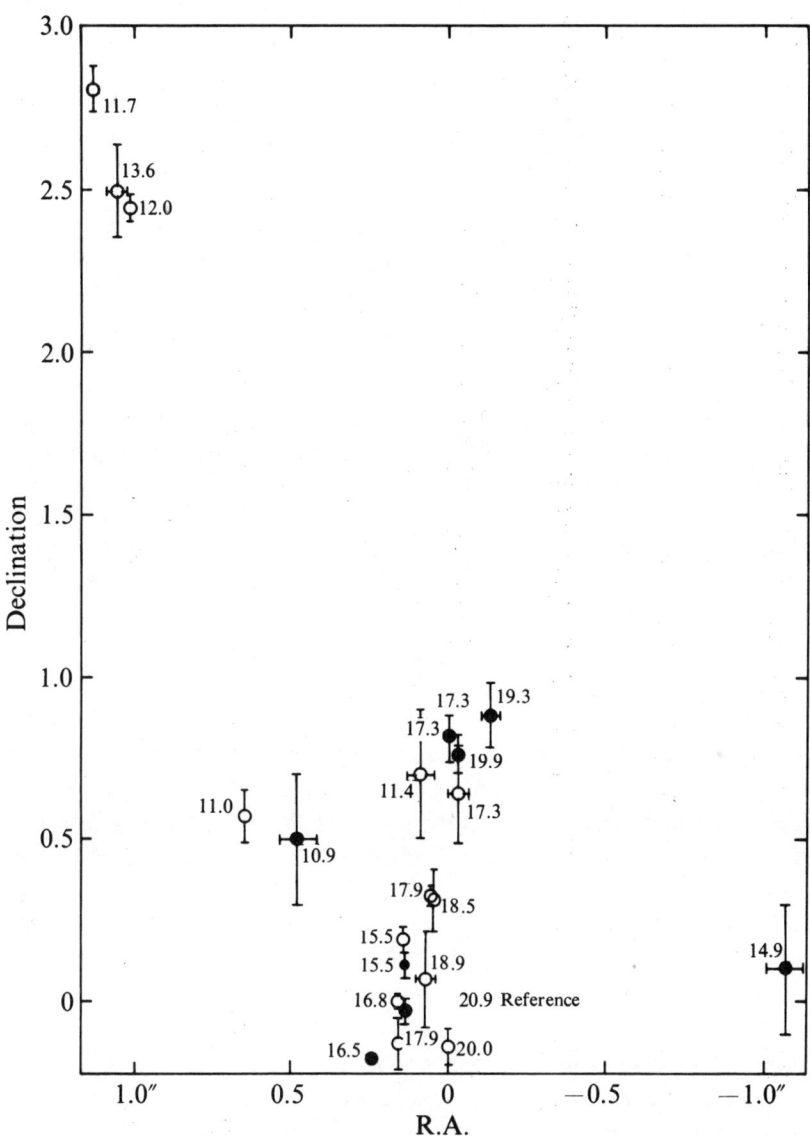

Figure 3.5. Map of the elementary sources in W49 (position 1) at 1665 MHz. The figures are the Doppler shifts of the components in km s^{-1} of RH ● polarization and LH ○ polarization. Positions are measured relative to the 20.9 km s^{-1} LH component (Harvey and others, 1974).

Observations of celestial masers

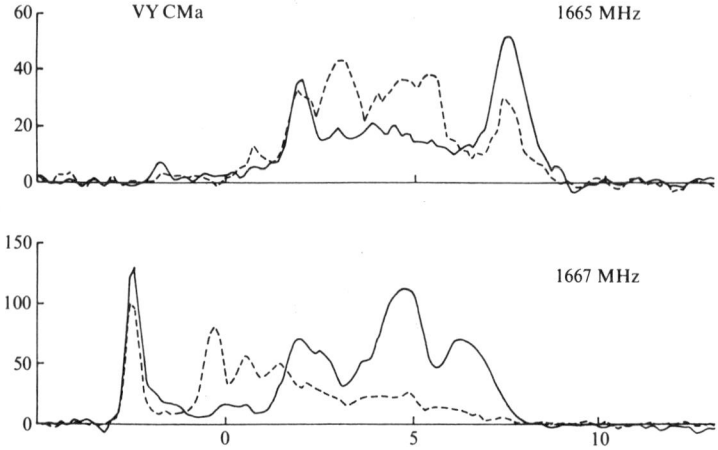

Figure 3.6. Spectra of VY CMa in LH (– – –) and RH (———) polarization at 1665 and 1667 MHz (Harvey and others, 1974). Abscissa: velocity in km s^{-1}. Ordinate: flux density in 10^{-26} Wm^{-2} Hz^{-1}.

great nebula in Orion is the most conspicuous) hidden from observations by eye or photography by clouds of dust and gas and are located by radio observations of thermal emission from the ionized hydrogen, the temperature of which is around 10 000 K. The catalogue label W often used for H-II regions (W3, W49) refers to the catalogue of thermal radio sources compiled by Westerhout (1958). The infra-red stars are thought to be clouds of hot dust and gas radiating prominently in the infra-red. The great majority of hydroxyl sources now known are close to either H-II regions or infra-red sources (ter Haar and Pelling, 1974). The water masers are similarly associated with H-II regions or with infra-red sources and, as a rough generalization, water masers occur together with hydroxyl masers although hydroxyl masers may occur without water masers (ter Haar and Pelling, 1974; Johnston, Sloanaker and Bologna, 1973).

3.2. Properties of selected sources

When early work (Weaver, Dieter and Williams, 1968) showed that hydroxyl maser sources were not resolved by single aerials, it was at once appreciated that interferometric observations would be needed for a full investigation of masers. The determination of the sizes of the sources of the individual components, mentioned above, requires interferometers with base lines of continental or intercontinental extent. In these so-called Very Long Base Line Interferometers (VLBI), signals are recorded separately on magnetic tape at the two receivers together with reference

3.2 Properties of selected sources

signals to give the relative phase, and the auto-correlation of the signals is then calculated in a computer (Moran and others, 1968; Burke and others, 1972; Moran and others, 1973). Examples of some results are given in table 3.1.

VLBI systems can resolve the individual sources, but to determine the relative positions of the sources, interferometers of shorter base-line are used (Cudaback, Read and Rougoor, 1966; Cooper, Davies and Booth, 1971; Harvey and others, 1974; Reisz and others, 1973). It is then possible to obtain the relative positions, frequencies (or line of sight velocities) and polarization of all the elementary sources that make up a complex source which is not resolved by a single aerial. Some results are shown in the form of maps in figures 3.5, 3.7 and 3.8. It is possible to

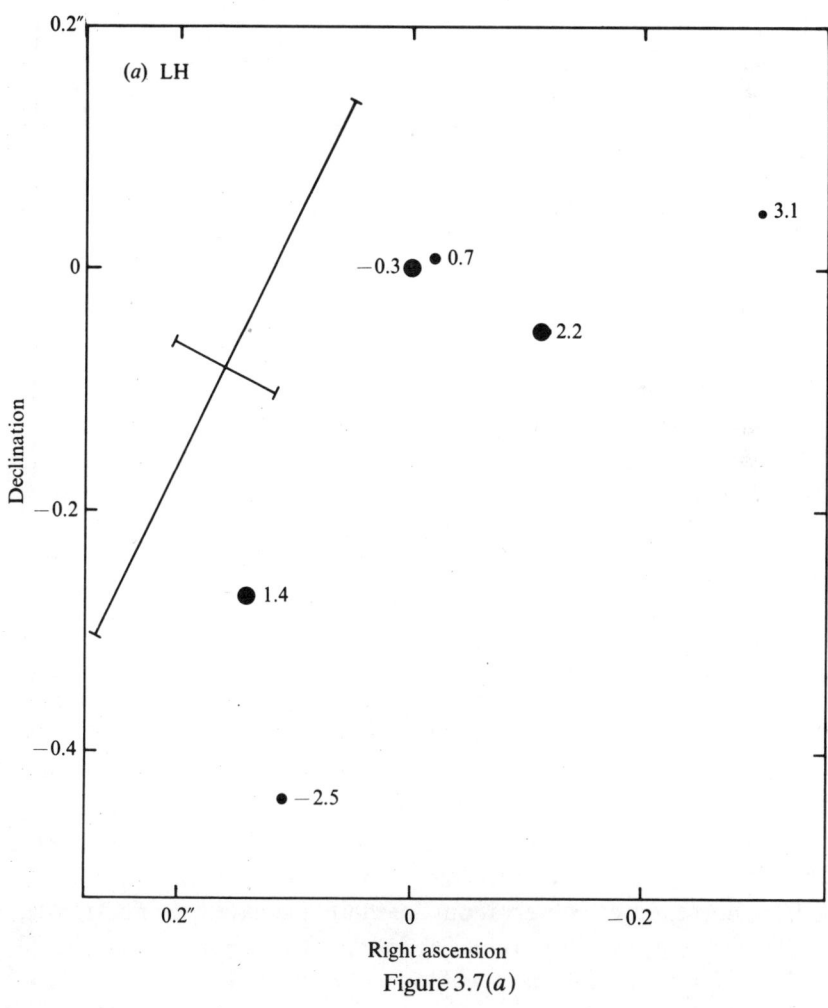

Figure 3.7(a)

Observations of celestial masers

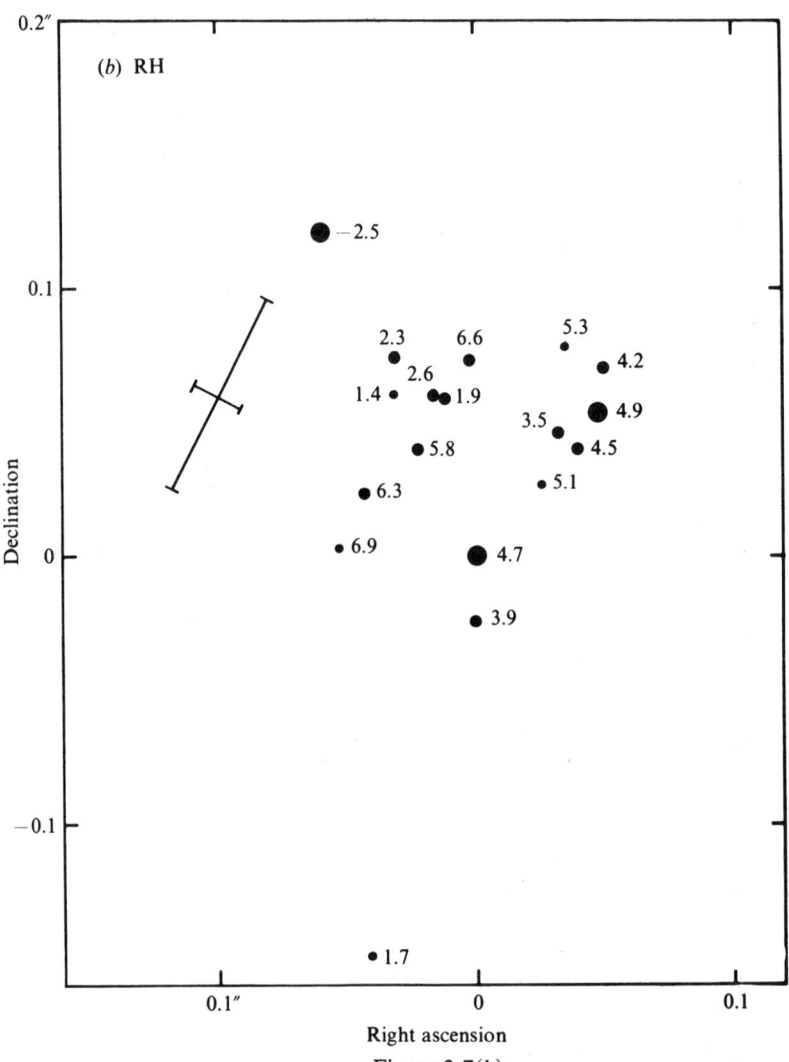

Figure 3.7(b)

Figure 3.7. Maps of (a) LH and (b) RH components in VY CMa at 1667 MHz (for description see figure 3.5). The origins of the two maps are unrelated (Harvey and others 1974).

observe at only one frequency at a time, and so the maps show the elementary sources which radiate at that frequency, 1665 MHz, for example. In only one instance (W3(OH)) have observations been made of the same source at two frequencies (Harvey and others, 1974).

The most extensive observations are those that have been made by R. D. Davies and his collaborators at Jodrell Bank (Harvey and others, 1974) using interferometers with spacings of 7×10^5 and 1.32×10^5

3.2 Properties of selected sources

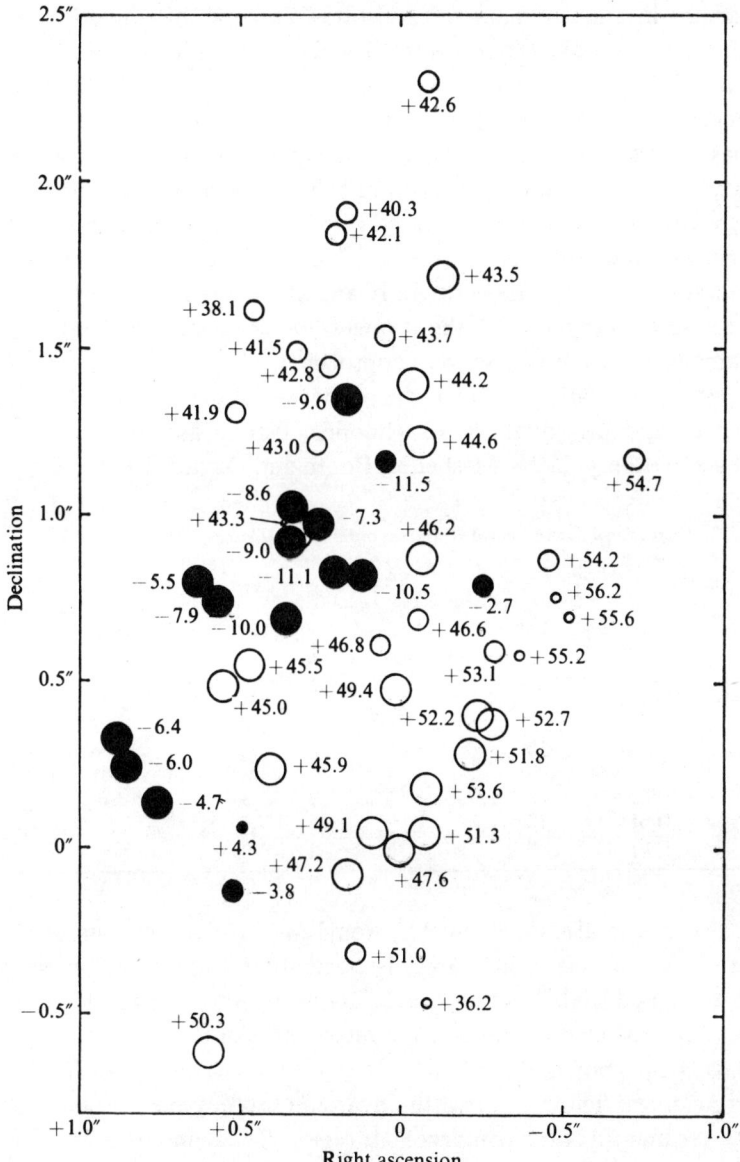

Figure 3.8. Map of components of VY CMa at 1612 MHz (for description see figure 3.5). Taken from Masheder and others (1974).

wavelengths operating at 1665 and 1612 MHz and, in the one case just mentioned, at 1667 MHz.

The observations on every hydroxyl source so far studied show the same picture of a number of elementary source regions, not more than

Observations of celestial masers

10^{-2} arcsec in angular diameter and often far less, distributed over a region a few seconds of arc across, each elementary source showing either left- or right-handed polarization and each radiating at a characteristic frequency corresponding to a relative mass motion of up to some 20 km s^{-1}. The radiation from an elementary source shows a spread of 1 or 2 parts in 10^6 in frequency. Davies and his collaborators have observed two types of source. In one, associated with H-II regions (figures 3.5, 3.7) the strongest radiation is at 1665 or 1667 MHz and the number of elementary sources may be between 10 and 25. In others, associated with infra-red sources (figure 3.6), the strongest radiation is at 1612 MHz and the number of elementary sources approaches 50.

Table 3.2 summarizes some of the properties of the sources observed by Davies and his collaborators (Cooper, Davies and Booth, 1971; Harvey and others, 1974; Masheder, Booth and Davies, 1974).

Table 3.2. *Properties of some hydroxyl sources observed by Harvey and others* (1974)

Source	Transition (MHz)	Number of components		Range of Doppler velocity (km s^{-1})	
		LH polarization	RH polarization	LH polarization	RH polarization
W3(OH)	1665	9	13	4.6	9.1
W49(1)	1665	14	8	9.9	9.0
W49(2)	1665	13	11	7.1	9.3
W75	1665	3	4	1.6	7.0
Sgr B2	1665	6	1	7.3	
VY CMa	1665	12		5.7	
	1667	6	18	5.6	9.4

While the general association of hydroxyl and water maser sources with H-II regions and infra-red sources is clear, there is only one object of which a detailed study has been made. Baldwin, Harris and Ryle (1973) made a detailed map of the thermal radio emission at 5 GHz from the compact H-II region in W3 (the densest part of the whole ionized region) and argued (see figure 3.9) that the hydroxyl sources most probably lie around the margin of the compact H-II region. The demonstration is not conclusive because the origins of direction of the two surveys (from Jodrell Bank and Cambridge) are not sufficiently well defined (a precision of about 10^{-1} arcsec would be needed) but when approximately aligned, the hydroxyl sources do all lie on the margins of the compact H-II region, and any other alignment would leave some outside the region and others apparently inside it.* Since hydroxyl is dissociated at the temperature of

---

* Two faint sources lie apparently within the H-II region.

3.2 Properties of selected sources

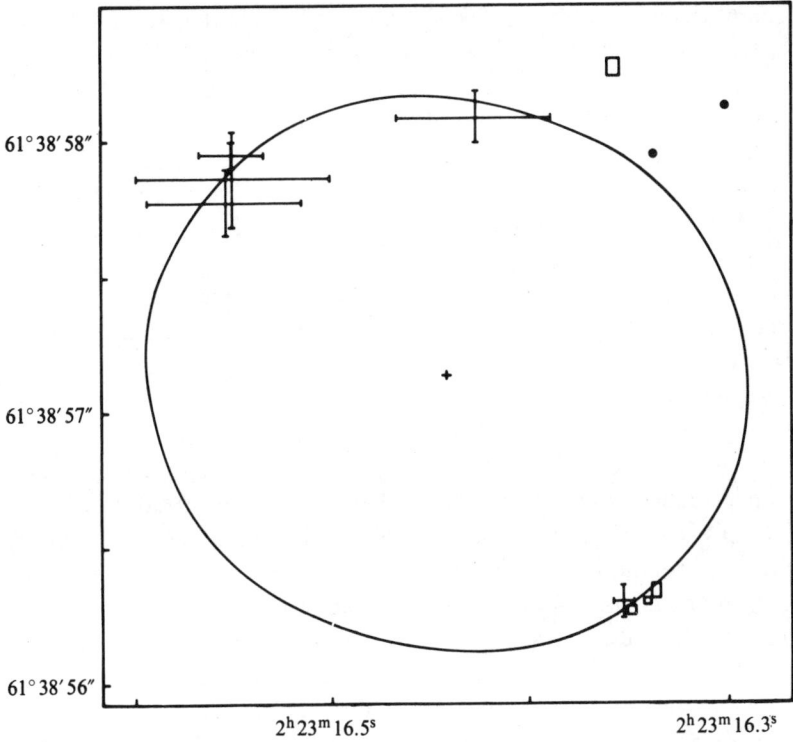

Figure 3.9. Superposition of components of W3(OH) at 1665 MHz on map of continuum radiation at 5 GHz (Baldwin and others, 1973).

the ionized hydrogen of an H-II region, the hydroxyl sources cannot lie physically within the H-II region.

The elementary sources associated with infra-red objects appear to occupy much more of the volume of the source region than do those associated with H-II regions, as may be seen, for example, by comparing figures 3.5 and 3.8. It may be reasonable to think of the maser sources associated with H-II regions lying in a thin shell between the ionized gas and un-ionized gas and dust around it, while the maser sources associated with infra-red objects lie in a relatively much thicker and cooler shell around the actual infra-red emitter.

The mass velocities of the sources associated with H-II regions show no obvious systematic arrangement, but those of sources associated with infra-red sources have been interpreted as corresponding to a rotation of the source as a whole, combined with an expansion of the radiating gas.

An important feature of all hydroxyl sources is that circularly polarized radiation with only one sense of polarization is emitted from any one elementary source (Cooper, Davies and Booth, 1971; Harvey and others, 1974). It follows that the polarization cannot arise simply from radiation in a magnetic field, for if it did so, radiation of both senses of polarization would be emitted from the same elementary source, at frequencies (more than two in general) displaced by the Zeeman shifts corresponding to the magnetic field. Maser sources have never been observed to radiate at more than one frequency and never in both senses of polarization from an individual elementary source.

The results of observations of sources at one frequency show that each direction in the sky is associated with a frequency shift which, it will be argued later (section 3.3), is the result of a Zeeman shift corresponding to an applied magnetic field and a Doppler shift corresponding to a mass velocity along the line of sight. Suppose then that each point in the source area may be associated with a frequency shift and conversely, that a distinct frequency shift means that radiation comes from a different part of the sky. In that case, it should be possible, by examining the frequency shifts of radiation from the four different transitions to determine whether any elementary sources radiate in more than one transition. If none of the frequency shifts in each of two transitions coincides, it is reasonable to suppose that none of the elementary sources radiates in more than one transition, but if some do coincide, then it may be that some sources radiate in more than one transition. The frequencies of the components of the spectra of the separate transitions in a number of well observed sources have been examined (Cook, 1975) and it has been argued that the evidence shows that elementary sources each radiate in only one transition. An example of the data is given in table 3.3.

The argument is indirect. It depends on associating a particular direction in the source with a particular shift of frequency as found in one transition, and then asserting that only if the frequency shift of a component of a second transition is identical with one of the first, might the same source region radiate at both frequencies. The argument must be indirect because, save in one case, no sources have been studied in sufficient detail at more than one frequency.

It is almost always found that the number of distinct sources (as identified by frequency shifts) is appreciably greater for one transition than for the three others in a hydroxyl source. Some statistics are listed in table 3.4.

When components can be separated by frequency, polarization and direction, the variation with time of each may be determined. Figure 3.10 shows the results that have been obtained for the hydroxyl source W3(OH) at 1665 MHz.

3.2 Properties of selected sources

Table 3.3. *Doppler shifts of emission lines in W49*

Doppler shift (km s^{-1})	1720	1667	1665	1612
+ 0.78			*	
+ 0.97		*		
+ 1.86			*	
+ 1.87		*		
+ 2.59		*		
+ 2.76			*	
+ 4.39		*		
+ 4.93		*		
+ 5.28			*	
+ 5.83		*		
+ 7.71			*	
+ 7.80		*		
+ 7.90	*			
+ 9.15			*	
+ 9.78			*	
+11.13			*	
+12.12			*	
+13.29		*		
+13.65			*	
+14.10				*
+14.50	*			
+14.85				*
+15.18			*	
+15.54		*		
+16.07		*		
+16.08			*	
+16.45				*
+16.79		*		
+16.80			*	
+18.06			*	
+19.04		*		
+19.10	*			
+19.49		*		
+19.50			*	
+20.04			*	
+20.50	*			
+20.94			*	
+21.20		*		
+22.02			*	

Observations of celestial masers

Table 3.4. *Numbers of emission components in different ground state transitions of hydroxyl sources (Weaver, Dieter and Williams, 1968)*

Source	Number of components			
	1612	1665	1667	1720
W3(OH)	2	11	4	3
W75	1	11		2
Ori A	3	16	1	
W51	1	6		3
NGC 6334		12	15	2
W49	4	19	16	4

More recent data (see table 3.2) usually enable more components to be identified in the stronger transitions.

3.3. The polarization of maser sources

The observation that elementary hydroxyl sources radiate in only one sense of circular polarization and not the complete set of Zeeman components corresponding to an applied magnetic field, is one of the pieces of evidence that show that amplification by stimulated emission is taking place. If a cloud of gas were emitting spontaneously, both senses of polarization would be observed, together with a spread of frequencies corresponding to a range of mass motions of the gas along the line of sight and possibly a range of magnetic fields. If radiation incident upon a molecule is to stimulate emission of a photon, the frequency of the incident radiation in the rest frame of the molecule must correspond to the frequency of the transition within the natural width of the transition. The natural width of the Λ-doublet transitions in hydroxyl, with lifetimes of 10^{12} s or so, are negligible; if, however, radiation is incident on an assembly of molecules in thermal motion, the frequency of the incident radiation will coincide with the transition frequency for some molecules provided the frequency of the incident radiation lies within the range of Doppler shifts corresponding to the thermal velocities of the molecules. The range of frequencies in elementary sources is about 1 to 2 kHz, or around 1 part in 10^6, corresponding to a kinetic temperature of some 50 K, and the equivalent range of velocities is about 0.3 km s^{-1}. In the absence of a magnetic field, the mass velocities along the line of sight should lie within the same range if radiation travelling along that direction is to be amplified. Now it is clear that much larger mass velocities, up to 20 km s^{-1} (as evidenced by the observed speed of frequencies) occur in the source region and so it is unlikely that velocities along an arbitrary line

3.3 The polarization of maser sources

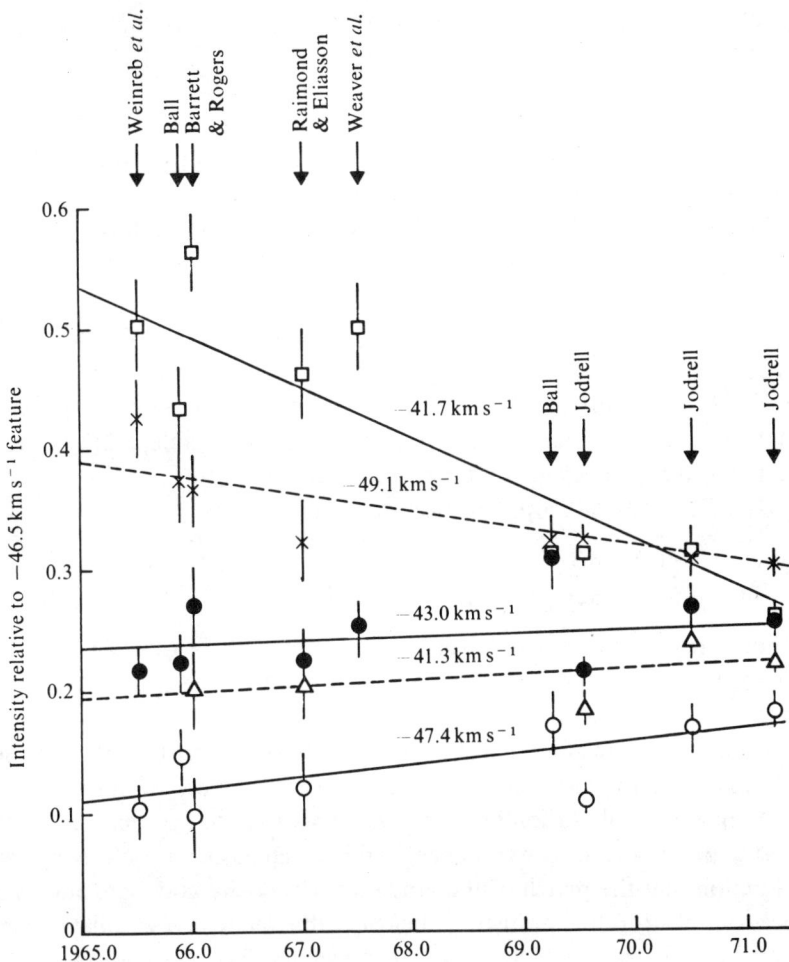

Figure 3.10. Variation of components of W3(OH) with time (Wilson, Davies and Ellder, 1972). The intensities of the −41.3, −41.7, −43.0, −47.4 and −49.1 km s^{-1} features are expressed as ratios to the intensity of the −46.5 km s^{-1} feature. × −49.1, LH+RH, ○ −47.4, LH+RH, ● −43.0, RH, □ −41.7, RH, △ −41.3, RH.

of sight will often agree to within a few tenths of a kilometre per second. That consideration may well suffice to account for the relatively small projected area of a source region from which radiation is observed. Now suppose that a magnetic field is applied to a gas, and suppose that both the field and the mass velocity change with distance x along the line of sight. Then the difference in Doppler shifts between two points separated by δx

will be

$$\frac{\nu}{c}\frac{\partial v}{\partial x}\delta x,$$

where ν is the frequency of the transition, c is the speed of light and v the component of mass velocity along the line of sight. Let B be the component of magnetic field along the line of sight. Then the changes in Zeeman shifts in a distance δx are

$$\pm \gamma_i \frac{\partial B}{\partial x} \cdot \delta x.$$

The values of the coefficients, γ_i, for the four transitions of the ground state Λ-doublet system in hydroxyl were listed in table 2.2(b) as the changes in frequency for unit magnetic induction.

The Zeeman and Doppler shifts will cancel if

$$\frac{\nu}{c}\frac{\partial v}{\partial x} = \gamma_i \frac{\partial B}{\partial x}.$$

There are two conditions for cancellation for a given value of γ_i – one gives left-handed and the other gives right-handed polarization, but only one can be satisfied at a time.

Evidently, if the gradients of field and velocity are constant along a line of sight, the Zeeman and Doppler shifts will cancel throughout the column and amplification by stimulated emission can take place.

It is an important consequence of this mechanism for polarizing the radiation that the match of the gradients of velocity and magnetic field occurs only for one transition because the factors $\nu/\gamma c$ differ from transition to transition as shown in table 3.5. The mechanism would therefore produce a filter which would suppress all amplification by stimulated emission except in the rare directions in which the velocities and magnetic fields match and then only in one transition and one sense of polarization. The frequency of the amplified radiation would be that given by the Doppler shift corresponding to the velocity at which the magnetic field is zero. Thus it may be supposed that directions in a source are characterized by the gas velocities at zero field, and it is these that are determined by the shifts of frequency of the observed radiation from the local standard.

The polarization mechanism just outlined (Cook, 1966, 1975; see also Shklovskii, 1969) has been criticized (Litvak, 1970) on the grounds that the necessary coincidence would be very unlikely. However, it is on just that basis that it is advanced, for the random alignment of mass velocities

3.4 Radiation from higher rotational states of hydroxyl

Table 3.5. *Values the factors $\nu/\gamma c$ for magnetic sub-levels of transitions in the ground state of hydroxyl*

Frequency (MHz)	$\nu/\gamma c$ (10^{-10} T s m^{-1})		
1612	±16.31	±5.47	±3.29
1665		±3.39 (mean of 2 close values)	
1667		±5.66 (mean of 4 close values)	
1720	±17.66	±5.85	±3.50

along an arbitrary line of sight would be expected to suppress maser action in most directions, and the chance of seeing maser radiation in an arbitrary direction is indeed small, about 1 in 10^5, as may be estimated from the ratio of the projected areas of individual sources to the total area of a source region.

The general nature of the argument that associates directions in the sky uniquely with mass velocities is however not affected if in contrast to the mechanism described above, it is supposed that the radiating molecules are in small, more or less spherical clumps (rather than spread out along a line of sight), each clump moving with a characteristic velocity. In such a model once again, a shift of frequency identifies an elementary source.

3.4. Radiation from higher rotational states of hydroxyl

It was seen in chapter 2 that the higher rotational states of the hydroxyl molecule exhibit Λ-doubling with hyperfine interaction just as does the ground state ($^2\Pi_{\frac{3}{2}}, J = \frac{3}{2}$) and the frequencies of the transitions within many of the systems have been measured. There is in each state a quartet of possible transitions, with the exception of the lowest state of the $^2\Pi_{\frac{1}{2}}$ system. When $J = \frac{1}{2}$, the possible values of F are 0 or 1 but the selection rules for electric dipole radiation do not allow the transition between the two sub-levels with $F = 0$. Thus of the four apparently possible transitions, three in fact occur (see table 2.3). The observations of radiation from the higher rotational states are summarized in table 3.6.

5 cm radiation from the $^2\Pi_{\frac{3}{2}}$, $J = \frac{5}{2}$ state has been observed from nine sources in the strong transitions at 6031 MHz ($F = 2 \rightarrow F = 2$) and 6035 MHz ($F = 3 \rightarrow F = 3$). The emissions from W75 and NML Cyg vary in time, that from NML Cyg showing variations within 12 hours (Zuckerman and others, 1969) (see figure 3.11).

Interferometric observations with long base lines show that the apparent sizes of sources in W3(OH) at a wavelength of 5 cm are the same as at 18 cm, indicating that the apparent sizes of the sources are not the result

Table 3.6. *Observations of radiation from higher rotational states of OH*

State	$^2\Pi_{\frac{3}{2}}, J=\frac{5}{2}$			$^2\Pi_{\frac{3}{2}}, J=\frac{7}{2}$	$^2\Pi_{\frac{1}{2}}, J=\frac{1}{2}$	
Transition, change of F	$3\to 3$	$2\to 2$	$4\to 4$	$1\to 0$	$0\to 1$	$1\to 1$
Source						
W3(OH)	1, 6, 11, 14, 17	1, 6, 11, 14, 17	13	2, 12, 15, 16		
W3(continuum)	9	9		2, 10, 15, 16		
W49 A	17			15, 16		
W51	18	18				
W75 N	11, 17					
Sgr B2	17, 18	18		2, 3, 9, 12	2, 3, 5, 7, 8, 9, 12	2, 3
ON-1	9			2		
NGC 7538	8					
NGC 6334 A	1, 4, 6, 17	17		12		
Orion A				8		
M17	18	18				
NML Cyg	17	17				
OH 69.5–1.0	18					

References
1. Ball and others, 1972; 2. Baudry, 1974; 3. Gardner and Ribes, 1971; 4. Gardner, Ribes and Goss, 1970; 5. Gardner, Ribes and Sinclair, 1971; 6. Knowles and others, 1973; 7. Palmer and Zuckerman, 1970; 8. Rickard, Zuckerman and Palmer, 1972; 9. Rickard, Zuckerman and Palmer, 1973; 10. Rydbeck and Ellder, 1973; 11. Rydbeck, Kolberg and Ellder, 1970; 12. Thacker, Wilson and Barrett, 1970; 13. Turner, Palmer and Zuckerman, 1970; 14. Yen and others, 1969; 15. Zuckerman and Palmer, 1970; 16. Zuckerman and others, 1968; 17. Zuckerman and others, 1972; 18. Rickard, Zuckerman and Palmer, 1975.

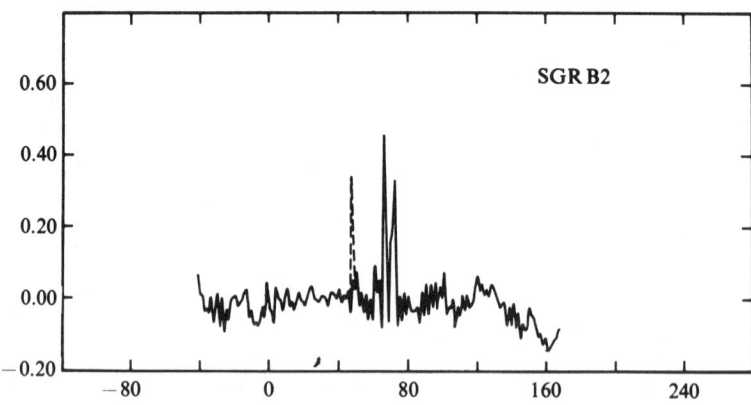

Figure 3.11. Spectrum of $^2\Pi_{\frac{3}{2}}, J=\frac{5}{2}, F=3\to 3$ transition at 6035 MHz in Sgr B2 (Rickard, Zuckerman and Palmer. © 1975. The University of Chicago). Continuous line: as on 1 Dec. Broken line: as on previous day. Abscissa: Doppler velocity km s^{-1}. Ordinate: aerial temperature.

3.4 Radiation from higher rotational states of hydroxyl

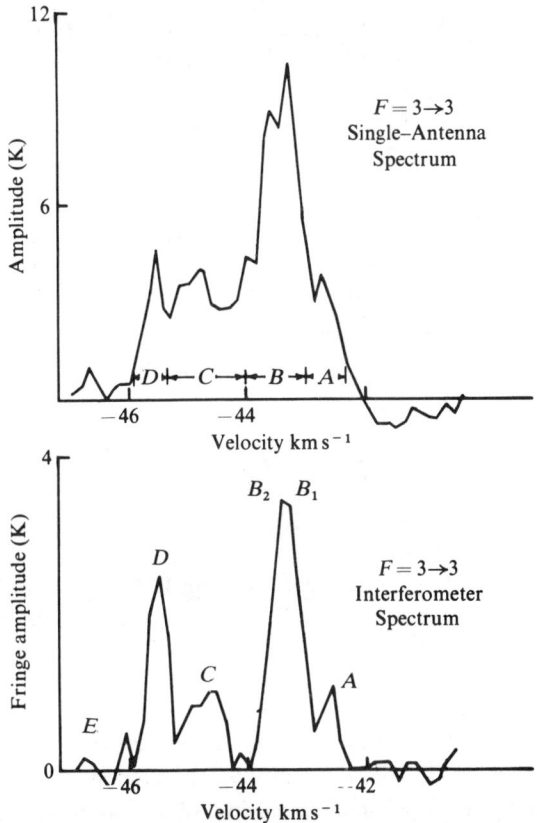

Figure 3.12. Spectrum of $^2\Pi_{\frac{3}{2}}$, $J = \frac{5}{2}$, $F = 3 \to 3$ transition at 6035 MHz in W3(OH) (Knowles, Johnston, Moran and Ball. © 1973. The University of Chicago). Features labelled $A, \ldots, E$. B is split into two polarized components, B_1 and B_2.

of interstellar scintillation (Knowles, Johnston, Moran and Ball, 1973). Figure 3.12 shows the spectrum of the $F = 3 \to 3$ 5 cm transition in W3(OH) and figure 3.13 is a map drawn from interferometric observations of the relative positions of the six components identified as A, B_1, B_2, C, D and E in figure 3.12. If the locations are compared with those of the positions of the 18 cm sources at 1665 MHz in W3(OH) (figure 3.9) it will be seen that the spread of components is similar, namely about 1 arcsec for the 5 cm radiation and 2 arcsec for the 18 cm radiation, that the range of velocities is similar, but that the 5 cm features fit only in a general way on to the 18 cm features. The three allowed transitions at 6 cm in the $^2\Pi_{\frac{1}{2}}$, $J = \frac{1}{2}$ state have been detected in seven sources and those from the $^2\Pi_{\frac{3}{2}}$, $J = \frac{7}{2}$ state have been detected only from W3(OH). Radiation from

51

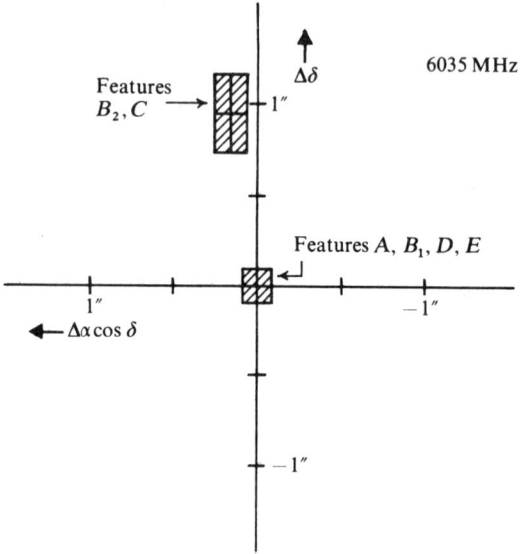

Fig. 3.13. Map of six components (see figure 3.12) of radiation from the $^2\Pi_{3/2}$, $J = 5/2$, $F = 3 \to 3$ transition in W3(OH) (Knowles, Johnston, Moran and Ball. © 1973. The University of Chicago).

the states $^2\Pi_{1/2}$, $J = 3/2$ and $J = 5/2$ has been sought without success. Rickard, Zuckerman and Palmer (1975) have summarized extensive observations of emission from higher rotational states. In addition to maser radiation they have detected broad quasi-thermal features in the spectra of W3(continuum) and Sgr B2 and failed to find them in six other regions, while they detected absorption by OH in the $F = 3 \to 3$ and $2 \to 2$ transitions of the $^2\Pi_{3/2}$, $J = 5/2$ rotational state in W3(continuum). Altogether they found 11 sources to show maser radiation from one or more excited rotational states and failed to detect $^2\Pi_{3/2}$, $J = 5/2$ radiation from 24 regions and $^2\Pi_{1/2}$, $J = 1/2$ radiation from 12 regions. There are some observations of polarized radiation from higher rotational states. Rydbeck, Kolberg and Ellder (1970) found that in W3(OH) the emission at 6035 and 6031 MHz were both circularly polarized, and that at 6035 MHz from W75 B had two circularly polarized components; the latter also showed variations correlated with those at 1667 MHz. Table 3.7 contains a summary of the numbers of components identified by Rickard, Zuckerman and Palmer (1975) in five cases and by Knowles, Johnston, Moran and Ball (1973) in two cases. More recently Knowles, Caswell and Goss (1976) in a survey of the $F = 3 \to 3$ (6035 MHz) transition of OH in the southern skies with the Parkes telescope in Australia, have detected a further 11 sources, one of

3.5 The distribution of hydroxyl sources

Table 3.7. *Number of distinct components for radiation from rotationally excited states of hydroxyl*

Source	State and transition $^2\Pi_{\frac{3}{2}}, J=\frac{5}{2}$	Number of components LH polarization	RH polarization	Reference
W3(OH)	$F = 3 \to 3$	6†		(a)
W49	$F = 3 \to 3$	1	3	(b)
W51	$F = 2 \to 2$	1	1	(b)
W51	$F = 3 \to 3$	3	4	(b)
NGC 6334 N	$F = 3 \to 3$	2†		(a)
NGC 7538	$F = 3 \to 3$	1		(b)
OH 69.5–1.0	$F = 3 \to 3$	2	2	(b)

† Polarization not distinguished.
References
(a) Knowles, Johnston, Moran and Ball, 1973.
(b) Rickard, Zuckerman and Palmer, 1975.

which in M17, appears to emit more photons at 5 cm than it does at 18 cm.

To summarize, it seems that the radiation from rotationally excited states shows similar properties to those of radiation from the ground state, and any differences may probably be accounted for by the fact that the intensity of the excited state radiation is normally much less than that from the ground state: thus one would expect to see fewer distinct components. The widths of excited state lines are for the most part about 1.5×10^{-6} of the centre frequency, much as for the ground state lines, implying that the kinetic temperatures are similar and that line-narrowing is not very significant. The range of velocities of the various components is also similar to that in the ground state.

3.5. The distribution, associations and statistics of hydroxyl sources

Stimulated emission from hydroxyl was first detected from the direction of H-II regions and for some time it was in the directions of other such regions that the search for further sources was concentrated. Subsequently attempts were made to carry out more systematic, less biased searches for sources throughout the galaxy. The first attempts to study the distribution and association of sources were undoubtedly affected by the bias of searches being conducted first in the direction of H-II regions and subsequently in those of infra-red sources. Turner (1969), however, did

Observations of celestial masers

attempt to minimize such bias and, on the whole, the classification of sources that he produced (1970) has accepted subsequent data. Lists of sources and summaries of properties have been given by Chaisson and Dickinson (1972), Dickinson and Turner (1972), Fillit and others (1972), Turner (1969, 1970), Wilson and others (1972) and Evans and others (1976).

Turner's (1970) classification is given in table 3.8. Primarily it classifies sources according to the strongest emission from the ground state Λ-doublet system. The primary distinction is between those sources which radiate most strongly at 1665 or 1667 MHz, and those which radiate most strongly at 1612 or 1720 MHz; the sources of the former group are associated with H-II regions, while the predominant association of the latter group is with infra-red stars.

Table 3.8. *Turner's classification of anomalous OH emission sources*

Property	Class I	Class IIa	Class IIb
Strongest emission line	1665 MHz (sometimes 1667 MHz)	1720 MHz	1612 MHz
Other emission lines	1667, 1612, 1720 MHz	None	1665, 1667 MHz
Absorption lines	None	1612, 1665, 1667 MHz	1720 MHz
Polarization of strongest lines	Circular (sometimes linear)	Practically none	Practically none
Polarization of other lines	Circular (sometimes linear)		Circular
Physical association	H-II regions	Supernova remnants (a few H-II regions)	Infra-red stars
Velocities	Agree with hydrogen recombination lines	Not associated with continuum	Usually two distinct velocity ranges
Relations between lines	None	All four have same velocity	1665 and 1667 MHz lines cover larger velocity range than 1612 MHz
Examples	W3, W49	W28, W44	NML Cyg, VY CMa
Numbers	50	8	26

The physical association of the hydroxyl group with an H-II region or infra-red star is initially based on a close agreement of the directions of the objects, but it is possible to be more discriminating with H-II regions because the Doppler shifts of the hydrogen recombination lines give the velocity of the H-II region along the line of sight and it is

3.5 The distribution of hydroxyl sources

almost always found that the spread of Doppler shifts of the hydroxyl sources includes the Doppler shift of the hydrogen recombination lines.

Turner's classification was originally based on some 50 sources; at least 137 are now known but most fit into Turner's scheme. A few sources do not fit well. Thus, there is a class I source apparently associated with a planetary nebula NGC 2348, and there are others associated with galactic nebulae that do not emit a strong radio continuum. There are a few sources associated with infra-red stars which emit more strongly at 1665 or 1667 MHz (see ter Haar and Pelling, 1974). Harris (1974) has shown that of six hydroxyl masers she studied, three of Turner's type I are associated with strong continuum radiation (about 80×10^{-29} W m^{-2} Hz^{-1}) at 5 GHz while the other three, of Turner's type IIb, are associated with much weaker radiation (less than 10×10^{-29} W m^{-2} Hz^{-1}) at 5 GHz. The 5 GHz radiation presumably comes from thermal sources. All three sources of type I are associated with water masers.

Turner's classification is based solely on the ground state radiation and, indeed, the few observations of radiation from higher rotational states are inadequate for helping to classify sources. So far, in fact, emission from higher states has only been observed from regions which give the most intense ground state emission, a few sources in Turner's class I and NML Cyg in his class IIb; Knowles, Caswell and Goss (1976) have observed that emission from the rotationally excited state $^2\Pi_{\frac{3}{2}}$, $J = \frac{5}{2}$, in a source of type I seems to be associated with the ground state radiation at 1720 MHz; apart from that no relation has been noted between the excitation of 'main' and 'satellite' lines in the ground and higher states. Indeed 'satellite' lines from higher rotational states seem to be relatively less frequent than from the ground state, although it is only quite recently that a substantial number of sources of radiation from excited states has been observed (Rickard, Zuckerman and Palmer, 1975; Knowles, Caswell and Goss, 1976).

The limitations of the overall picture of hydroxyl sources need to be emphasized. In many instances, sources have been sought where they were expected to be found – in association with H-II regions and infra-red objects. Sources are always associated, it seems, with one or other such objects, but no systematic estimate has been made of the proportion of H–II regions for example, with which no hydroxyl emission is associated (see Winnberg and others, 1973). It should also be noticed that some of the sources in Turner's class I have been observed only at 1665 or 1667 MHz, so that it is not known how well they fit his classification.

Observations of celestial masers

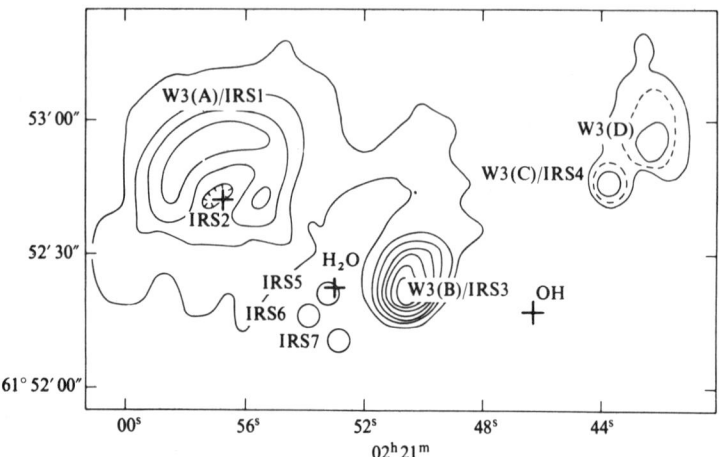

Figure 3.14. Map of various objects associated with the W3 thermal source. IRS1–7: Infra-red sources, A–D: Thermal radio sources (H-II regions), H$_2$O: Water source, OH: Hydroxyl source. Contours are of radio intensity. (Wynn-Williams, Weaver and Wilson, 1974.)

Lastly, emission from higher rotational states has been sought only in the neighbourhood of ground state emission.

Water and hydroxyl masers are not the only objects associated with the thermal continuum sources identified as H-II regions. As may be seen from the map of the W3 region in figure 3.14, not only are there at least two continuum sources and water and OH masers, but also infra-red sources. Other molecules have been detected in the region. The great nebula in Orion, another region which has been studied intensively, shows similar associations.

3.6. The power of hydroxyl sources

Most models of amplification by stimulated emission lead to the conclusion that the radiation is emitted in a restricted beam, so that an observer on Earth detects only the sources directed towards him. Suppose that $\delta\Omega$ is the solid angle embraced by a typical radiated beam. Then the chance of detecting a source from an arbitrary position is $\delta\Omega/4\pi$, and if the number of sources seen from the Earth in a given region is N, the total number of sources that radiate in some directions would be $4\pi N/\delta\Omega$.

The total power emitted from all sources in a given region (for example, the neighbourhood of W3(OH) or W49) can now be estimated without knowing $\delta\Omega$, although to find the total of elementary sources, $\delta\Omega$ must be known.

3.7 Water masers

Let the elementary sources have projected diameters d, so that the projected area of a source is $\frac{1}{4}\pi d^2$. The source radiates over a bandwidth $\delta\nu$ into a solid angle $\delta\Omega$. If the mean intensity of the radiation over the bandwidth is I, the power radiated by the source is

$$\tfrac{1}{4}\pi d^2 I \delta\nu \delta\Omega$$

and the power radiated by all sources in the region is

$$\frac{4\pi N}{\delta\Omega} \times \frac{\pi d^2}{4} I \delta\nu \delta\Omega,$$

that is

$$N\pi^2 d^2 \delta\nu,$$

a result that is independent of $\delta\Omega$ because the power radiated by a single source is proportional to $\delta\Omega$ whereas the total number of sources in a region is inversely proportional to $\delta\Omega$.

By a similar statistical argument, the ratio $\delta\Omega/4\pi$ may be seen to be of the order of the ratio of the projected cross-sectional area of a single source to that of the source region as a whole, namely about 1 in 10^6.

Table 3.9 contains estimates of the total power radiated by some sources for which sufficient detail is known; the last column gives the rate of emission of photons, a quantity relevant to some possible photon pumping mechanisms (see chapter 5).

3.7. Water masers

It was explained in chapter 1 that only one transition in the water molecule has been found from galactic sources, namely that at 22.2 GHz (1.35 cm) between the rotational levels 6_{16} and 5_{23}. Since the first detection by Cheung and others (1969) at least 60 sources have been found, including those in the direction of Sgr B2, W49 and Orion. Interferometers with very long base lines show that the angular sizes of water sources are of the order of 10^{-3} arcsec across or less, compared with 10^{-2} arcsec or less, typical of hydroxyl sources.

The characteristic features of water sources are very high intensity (the *aerial* temperature in W49 reaches 2000 K whereas the highest aerial temperature of an hydroxyl source is some 50 K), the small angular size and very high brightness temperature (10^{15} K for W49), the very large spread of frequencies, corresponding to a range of 500 km s^{-1} in W49 (see Sullivan, 1971; Goss and others, 1976), and the rapid variations in time. Unlike the hydroxyl radiation, water radiation is not strongly polarized, and when it is, it is linearly polarized (Cheung and others,

Table 3.9. *Total power emitted by maser sources*

(a) Hydroxyl masers (data from Harvey and others, 1974)

Source	Transition	Power (10^{21} W)	Photon rate (10^{45} s^{-1})
W3(OH)	1665	2.1	2
Sgr B2	1665	11	10
W49(1)	1665	54	50
W49(2)	1665	36	33
W75 S	1665	0.12	0.1
W75 N	1665	0.11	0.1
VY CMa	1665	0.35	0.3
	1667	0.84	0.8
	1612	5.4	0.5

(b) Water masers (data from Moran and others, 1973)

Source	Power (10^{23} W)	Photon rate (10^{45} s^{-1})
W49	1500	10^4
W3(OH)	3.5	24
Ori A	0.3	2
VY CMa	0.3	2

W49 is much the most powerful source.

1969; Buhl and others, 1969; Knowles and others, 1969; Meeks and others, 1969; Turner, Buhl, Churchwell, Mezger and Snyder, 1970). The absence of circular polarization is without doubt a consequence of the very small Zeeman splitting in water, about 10^{-3} of that in the Λ-doublet of hydroxyl. In consequence the circularly polarized components would not have significantly different frequencies in magnetic fields in which those of hydroxyl would be clearly separated; if Zeeman splitting occurred, the two components of opposite sense would be superposed giving apparently linear polarization.

The water source studied in most detail is W49. Some of its properties have just been mentioned. Johnston and others (1971) have determined interferometrically the directions of components with small shifts of frequency (low velocity components) and find that, as with hydroxyl radiation, each elementary source at a distinct position is characterized by a distinct frequency (or mass velocity): figure 3.15.

3.7 Water masers

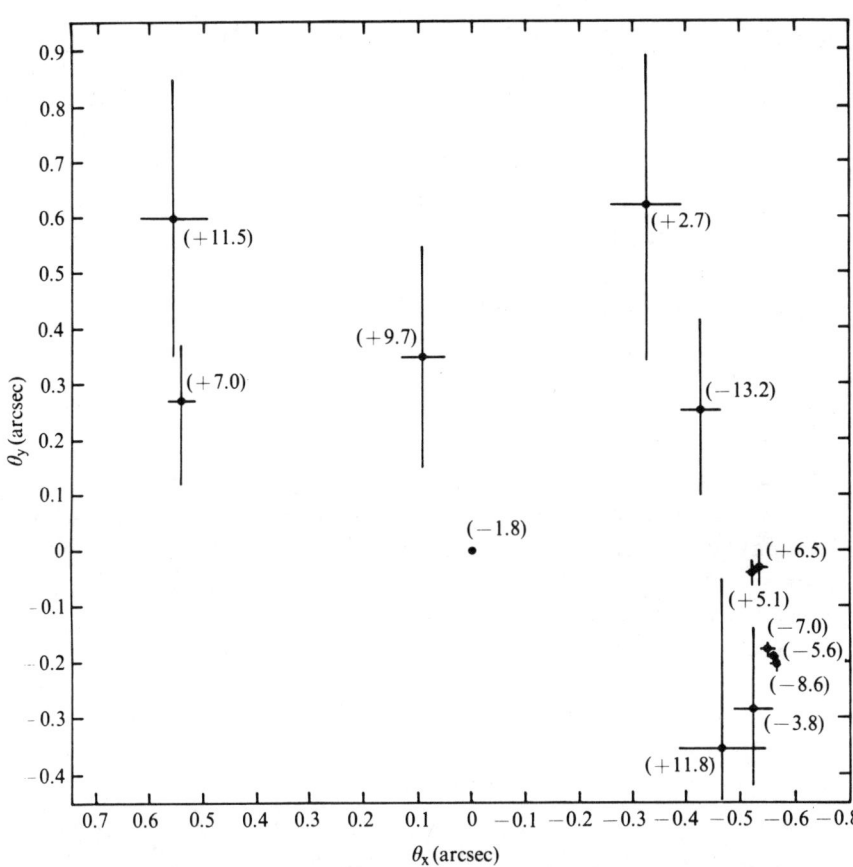

Figure 3.15. Map of components of the water source in W49, 1970 June 21 (Johnston and others. © 1971. The University of Chicago). Figures are Doppler shifts in km s^{-1}.

Some components of W49 change their intensities very rapidly: the 15 km s^{-1} and 9 km s^{-1} components varied by more than twice in the one month from 1969 Jan 9 to Feb 8 (see also figure 3.4).

The flux from W49 is some 4000 fu† and the equivalent brightness temperature, as has been mentioned, is about 10^{15} K.

Two water sources are found in the region of W3, one in the direction of the radio continuum source W3 itself (which also shows hydroxyl radiation at 1720 MHz), where a water source with a number of components is found, and one in the direction of the hydroxyl source W3(OH), with a main component at -49 km s^{-1} (relative to the local standard of rest), the size of which is less than 10^{-2} arcsec.

† 1 fu (flux unit) = 10^{-26} W ster^{-1} m^{-2} Hz^{-1}.

VYCMa is an infra-red object which shows hydroxyl maser radiation at 1612, 1665 and 1667 MHz, the frequencies corresponding to velocities over the range of -10 to $+55$ km s^{-1}; the equivalent velocity for the water radiation is $+18$ km s^{-1}. Nine other examples of water radiation from infra-red objects are known, the intensity in general being some 100 times less than that of water sources associated with H-II regions. Some water masers are associated with late M-type stars and have the particularly interesting property that the intensity of water radiation shows periodic variations in time that are correlated with those of the infra-red and optical radiation from the star (Schwartz, Harvey and Barrett, 1974); this is strong evidence for a physical association.

With radiation observed from only one transition, the material does not exist to set up a classification of water sources on the lines of Turner's classification of hydroxyl sources. As with hydroxyl sources, water sources have been sought where they might be expected to be found; with that limitation on sampling, water sources are in general associated with hydroxyl sources – but not invariably for there are many hydroxyl sources with no associated water sources, while there are some water sources not associated with hydroxyl radiation. In particular, whereas all hydroxyl sources so far detected lie in the disc of the galaxy, water sources have been found that are associated with cool M-type stars out of the disc of the galaxy (Johnston and others, 1973). Furthermore, in two cases, W49 and W3(OH), it has been shown that the hydroxyl and water masers are separated by at least 2×10^4 A.U. (Mader and others, 1975) and in a third case, W3, by 40 arcsec (Wynn-Williams, Weaver and Wilson, 1974). The association of water and hydroxyl sources has been discussed most recently by Yngvesson and others (1975).

3.8. Other masers

Intense radiation thought to come from maser action has been detected from other molecules besides water and hydroxyl. No radiation so intense had been detected until recently, and time variations and polarization have not been studied. Identification of maser action usually depends on a discrepancy between the brightness temperature and a significantly lower kinetic temperature. CH has a similar structure of OH and Λ-doublet radiation was sought without success for some time. Because of difficulties with experimental work, the frequencies were not well known, depending in fact on optical spectroscopy, but radiation was eventually detected (Rydbeck, Ellder and Irvine, 1973; Turner and Zuckerman, 1974).

3.8 Other masers

Rotational transitions in methyl alcohol and silicon monoxide are also considered to show maser action (Barrett and others, 1971; Chiu and others, 1974; Turner and others, 1972; Zuckerman and others, 1972). In particular, Hills, Pankonin and Landecker (1975) found that the methanol radiation in Orion comes from a cluster of small sources with a structure similar to that of water and hydroxyl sources. The Orion source seems to be associated with the water, hydroxyl and infra-red sources. Snyder and Buhl (1974) have in addition found strong unidentified radiation from the Orion complex at 3.48 mm and consider it may come from maser action.

Radiation from silicon monoxide is observed from rotational transitions in the vibrational levels $v = 0$, 1 and 2, the most intense radiation coming from $v = 1$, $J = 1 \to 0$ and $2 \to 1$, and $v = 2$, $J = 1 \to 0$. The vibrational levels lie at $1231\ \text{cm}^{-1}$ ($v = 1$) and $2449\ \text{cm}^{-1}$ ($v = 2$) above the ground state; the fact that they have appreciable populations shows that there are substantial departures from thermodynamic equilibrium. The radiation from silicon monoxide is intense, the flux in some sources approaching that from strong water sources (see table 3.10).

Table 3.10. *Observations of radio emission from silicon monoxide* (Pelling, 1975)

Frequency (GHz)	Line Transition	Sources		Brightness temperatures
43	$v = 1$, $J = 1 \to 0$	Ori A		2–60 K
		W Hya		
		VY CMa		
		R Leo		
		o Ceti		
86	$v = 1$, $J = 2 \to 1$	Ori A	R Leo	2–20 K
		U Ori	VX Sgr	
		RX Boo	X Cyg	
		W Hya	NML Cyg	
		U Her	S Per	
			R Cas	
129	$v = 1$, $J = 3 \to 2$	Ori A		2 K
43	$v = 2$, $J = 1 \to 0$	W Hya	o Ceti	1–30 K
		VY CMa	Ori A	
		R Leo		

4
Theory of amplification by stimulated emission

4.1. Principles

A theory of amplification by stimulated emission should if possible account for the exceptionally high intensities of the hydroxyl and water sources and the polarization and apparent sizes of the sources, and should also help to understand the variations in time and perhaps the fact that radiation from only one transition is observed from any elementary source spot. It should also be possible to use the theory to interpret the observations in terms of such properties of the sources as the density of hydroxyl or water molecules, density and temperature of other constituents of the source, including radiation, other molecules and electrons. At present none of these aims can be achieved in detail although the solutions to some can be sketched.

The theory explained in this chapter is that of *amplification* by stimulated emission. Laboratory masers and lasers incorporate a resonant structure and are self-oscillating; there is no resonant structure in astrophysical masers† and the radiation observed is amplified from some incident radiation. An important difference between astrophysical and laboratory masers is in consequence that the width of the emitted line is not extremely small but is of the same order as the Doppler width of spontaneous emission.

There seem, in general, to be three possible sources of the radiation. In the first place, there is spontaneous emission from the gas in the source itself. The radiation from a self-oscillating laboratory maser or laser may be thought of in the same sense as amplified spontaneous emission but in systems with resonant structures, the radiation in effect passes through a volume of amplifying medium a large number of times instead of just the once in an amplifier without a resonant structure. Although, because the column density of excited atoms or molecules that can usually be achieved in the laboratory is low, it is necessary to have a resonant system to obtain net gain by stimulated emission, yet there are some atoms or molecules which can be obtained in sufficiently high densities in the

---
† A microwave resonant structure has been postulated in astrophysical masers but is generally discounted and does not seem to be needed.

4.1 Principles

excited state in laboratory systems to allow amplification by stimulated emission without a resonant structure and a considerable amount of work has been done on the theory and experimental study of such systems in the laboratory (Allen and Peters, 1972; Peters and Allen, 1972a,b).

The second source is the cosmic background radiation, black body radiation corresponding to a temperature of 2.7 K; it will be amplified in the absence of any other source. Thirdly, there may be radiation from an associated thermal source, from the H-II region or the infra-red star within the shell of maser sources; it may be thought to be the most likely of the three possible sources and the correlation between infra-red and water radiation from late M-type stars (chapter 3) may arise from it.

Consider a molecule with two possible states, labelled i and j, supposed to be of equal statistical weights, the difference of energy between them corresponding to a frequency ν_{ij}. Then if the molecule is in thermal equilibrium with a radiation field of density u, the number of molecules in the upper state (i, say) is related to the number in the lower state by

$$n_i = n_j \exp\{-h\nu_{ij}/kT\},$$

where T is the temperature and k is Boltzmann's constant.

Further, the energy density of the radiation is given by

$$u = \frac{2h\nu^2}{c^3}(e^{h\nu/kT}-1)^{-1}.$$

If A_{ij} is the probability of a spontaneous transition from state i to state j and if the coefficients of stimulated emission are B_{ij} for downwards transitions and B_{ji} for upwards transitions the condition for equilibrium is that

$$B_{ji}n_j u = B_{ij}n_i u + A_{ij}n_i.$$

On substituting for u and n_i/n_j, the values for thermodynamic equilibrium, it follows that

$$B_{ji} = B_{ij} = \frac{c^3}{2h\nu_{ij}^3}A_{ij}.$$

If the statistical weights, g, of the states, namely $2F+1$, are unequal, then

$$g_j B_{ji} = g_i B_{ij}$$

and

$$B_{ij} = \frac{c^3}{2h\nu_{ij}^3}A_{ij}.$$

Amplification by stimulated emission

The intensity of the radiation and the probabilities A_{ij}, B_{ij}, B_{ji} are all functions of frequency. The transition probabilities for an isolated molecule have a Lorentzian distribution with a half width determined by the lifetime; the lifetimes of the excited states of the Λ-doublets of hydroxyl are so long that the widths of the Lorentzian distributions are negligible. In practice, of course, the molecules are not isolated and we have to consider groups of molecules moving with random thermal velocities about mean mass velocities. Emission or absorption by molecules will only occur at a particular frequency if the velocity of the molecules in a particular frame of reference is such that the corresponding Doppler shift of the frequency is equal to the difference of the frequency from the rest frequency in the frame. The intensity of the radiation emitted or absorbed will then be proportional to the number of molecules having that velocity, that is, it will be proportional to $\phi(\nu - \nu_0)\,d\nu$ where ϕ describes the distribution of molecular velocities and ν_0 is the frequency corresponding to zero velocity in a local standard of rest. ϕ is normalized so that $\int_0^\infty \phi\,d\nu = 1$.

Now consider radiation of intensity I travelling in the direction of increasing x and incident on a layer of molecules of thickness δx. Suppose the radiation to be confined to a solid angle $\delta\Omega$ about the direction of x. The density of the radiation will accordingly be $I\delta\Omega/c$. Molecules will absorb radiation from the beam at the rate

$$h\nu\phi(\nu)B_{ji}n_j(I\delta\Omega/c)\,\delta\nu\,\delta x$$

per unit area.

n_j is the number of molecules per unit volume in the lower state j.
The constant ν_0 has been omitted for brevity.
Similarly, radiation will be emitted at the stimulated rate

$$h\nu\phi(\nu)B_{ij}n_i(I\delta\Omega/c)\,\delta\nu\,\delta x.$$

The radiation remaining after absorption is coherent with the incident radiation, so that the absorbed radiation may be thought of as coherent with the incident radiation. Likewise the stimulated radiation is coherent with the incident radiation; in particular both absorbed and stimulated radiation are confined to the same solid angle $\delta\Omega$.

At the same time, the molecules emit spontaneous radiation incoherently and isotropically; the power radiated into the solid angle $\delta\Omega$ in the x-direction is

$$h\nu\phi(\nu)A_{ij}n_i(\delta\Omega/4\pi)\,\delta\nu\,\delta x.$$

4.1 Principles

If quantities are constant in time, the change of intensity in the distance δx is formally

$$\frac{\partial I}{\partial x} \cdot \delta\nu\,\delta x$$

and is equal to

$$h\nu\,\delta\Omega\frac{I_\nu}{c}\{B_{ij}n_i - B_{ji}n_j\}\phi(\nu)\,\delta\nu\,\delta x + h\nu\frac{\delta\Omega}{4\pi}A_{ij}n_i\phi(\nu)\,\delta\nu\,\delta x.$$

Now integrate with respect to frequency over the range of frequency within which radiation occurs, and define the equivalent width of the radiation, $\Delta\nu$, by

$$\Delta\nu = \frac{\int I\phi\,d\nu}{\int I\,d\nu}.$$

Then

$$\frac{\partial I}{\partial x} = (b_{ij}n_i - b_{ji}n_i)I + a_{ij}n_i,$$

where

$$b_{ij} = \frac{h\nu B_{ij}}{c\Delta\nu},$$

$$b_{ij} = \frac{h\nu B_{ji}}{c\Delta\nu},$$

and

$$a_{ij} = \frac{h\nu A_{ij}}{4\pi\Delta\nu}.$$

It should be noted that this procedure, which has often been adopted, ignores any possible difference between the profiles of the incident, stimulated and absorbed radiation and may be inadequate.

If, in the general case where the statistical weights are unequal, we let

$$\delta n_{ij} = g_j n_i - g_i n_j,$$

be the effective difference of molecule densities, then the equation for the rate of change of intensity in the direction x is

$$\frac{\partial I}{\partial x} = Ib_{ij}\,\delta n_{ij} + a_{ij}n_i.$$

65

Amplification by stimulated emission

So far it has been assumed that the intensity is not dependent on time. If it does depend on time, the radiation incident on a layer of gas at position x and time t has the intensity $I(x, t)$, while that emerging at $x + \delta x$ would have the intensity $I(x, t + \delta x/c)$ if there were no radiation from the layer. The total change in the distance x is thus

$$\frac{\partial I}{\partial t}\frac{\delta x}{c} + \frac{\partial I}{\partial x}\delta x,$$

Accordingly

$$\frac{\partial I}{\partial x} + \frac{1}{c}\frac{\partial I}{\partial t} = \frac{1}{c}\frac{dI}{dt} = Ib_{ij}\,\delta n_{ij} + a_{ij}n_i.$$

This is the fundamental form of the equation of radiative transfer for maser amplification. It could be solved in simple geometries if n_i and n_j were known. In general however, the numbers of molecules in the two states depend on u, the energy density, because for every photon emitted or absorbed, a molecule makes a transition from the upper to the lower state or *vice versa*. Thus, for an emission of a photon

$$\dot{n}_2 = -\dot{n}_1 = -uB_{ij}n_i.$$

In the special case, to be discussed below, where radiation travels in two streams I^+ and I^-, in the directions of x increasing and decreasing respectively,

$$u = \frac{1}{c}(I^+ + I^-)\,\delta\Omega.$$

Taking into account absorption and spontaneous emission the net gains or losses from the two levels are

$$\dot{n}_i = -\dot{n}_j = -uB_{ij}\delta n_{ij} - A_{ij}n_i$$

where, as before, $\delta n_{ij} = g_j n_i - g_i n_j$.

There are in addition other processes which change the numbers of molecules in the two levels. In the first place, each level may lose molecules by decay to yet lower states or by destruction of the molecules; thus

$$\dot{n}_i = -\gamma_i n_i,$$

$$\dot{n}_j = -\gamma_j n_j.$$

It seems likely that many masers are bathed in an infra-red or ultraviolet radiation field. If the field is isotropic or nearly so, the density may

4.1 Principles

be comparable with that of maser radiation confined to a narrow beam, and then transitions stimulated by the isotropic field will be important.

There must also be processes that add molecules to the upper state. Some may come from higher states of the molecule or by creation of molecules from atoms or other molecules at a rate

$$\dot{n}_i = p,$$

where p is independent of n_i and n_j. Some may come from the lower states of the Λ-doublet at rates proportional to n_j, so that

$$\dot{n}_i = -\dot{n}_j = p'n_j,$$

where p' is independent of n_i and n_j. The total rates are then

$$\dot{n}_i = -uB_{ij}\,\delta n_{ij} - A_{ij}n_i - \gamma_i n_i + p + p'n_j,$$
$$\dot{n}_j = uB_{ij}\,\delta n_{ij} + A_{ij}n_i - \gamma_j n_j - p'n_j.$$

The pumping rates p and coefficients p' in these equations have been taken to be independent of the number densities of the populations of the levels coupled by the maser transitions. Whether that is a good assumption or not depends upon the nature of the pumping process. Suppose, for example, collisions with some projectile excite molecules from a lower state to an upper state at a rate $p'n$ (lower). Clearly p' is not dependent on n (lower). Again, suppose a chemical process produces molecules in an upper state at a rate p. The rate will depend on the density of molecules involved in the chemical process but not upon the numbers of molecules in the levels connected by the stimulated transition. Pumping by radiative processes requires more consideration. Suppose a gas has 3 levels, 1, 2 and 3, 3 being the highest and the stimulated emission being from 2 to 1. Neglecting collisions, the populations in 1 and 2 will be inverted if the spontaneous rate $3 \to 2$ is greater than $3 \to 1$. The actual rates at which molecules enter levels 2 and 1 will however depend on the number in 3 which will depend on the radiation densities at the frequencies of the transitions $1 \to 3$ and $2 \to 3$ as well as on the numbers in levels 1 and 2. However, given that the pumping radiation cannot have a Planckian distribution (for then no inversion would occur) the radiation densities will be the consequence of radiative transfer in the gas and thus will depend on the densities of molecules in the three levels throughout the gas, being in that way coupled to the maser transitions.

The situation may however be simple in some models of maser sources. The isolated directions in which maser radiation is observed may be interpreted in either of two ways – the masers themselves are very few in number but radiate isotropically, or they are numerous but radiate only into narrow cones so that few are seen from the Earth. If the latter holds, the general region within which the masers lie will be filled by a large number of masers radiating independently at different frequencies in random directions. The flux of pumping radiation in the general region, in so far as it depends on the number of densities in the levels connecting by the masing transition, will depend on an average distribution of the maser radiation, but not directly on the radiation from any individual maser. Thus the pumping rate in the individual maser will be the product of the number densities in the maser column, with a probability of stimulated transition proportional to the density of pumping radiation which is independent of the number densities in the particular maser.

If, however, the masers are few in number and radiate isotropically, the foregoing simplification may not be justified.

Consider now the orders of magnitude of the terms in the equation of radiative transfer and in the rate equations.

The ratio of the stimulated to the spontaneous emission in the equation of transfer is

$$\frac{b_{ij}}{a_{ij}} \frac{\delta n_{ij}}{n_i} I.$$

Now $b_{ij}/a_{ij} = c^2/2h\nu^3$, and so the ratio is of order

$$\frac{n_i}{\delta n_{ij}} T_B,$$

where T_B is the temperature of a black body radiating at the same intensity. Because T_B is so very large (10^{12}–10^{14} K) the stimulated emission dominates the equation of transfer throughout most of a typical source, even though $\delta n/n$ may be no more than 10^{-3}.

Thus, in many cases, it will be sufficient to ignore the spontaneous term in the equation of transfer, and to take the equation as

$$\frac{\partial I}{\partial x} + \frac{1}{c} \frac{\partial I}{\partial t} = KI,$$

where $K = b_{ij} \delta n_{ij}$.

In the rate equations, on the other hand, the ratio of spontaneous to stimulated terms becomes

$$\frac{n_i}{\delta n_{ij}} \frac{4\pi}{\delta \Omega} \frac{h\nu^3}{Ic^2};$$

$\delta \Omega$ is the solid angle into which the stimulated emission radiates, and enters the ratio because the spontaneous change in numbers comes from

4.2 Unsaturated linear masers

spontaneous emission into the complete sphere of solid angle 4π, whereas the stimulated emission is confined to the solid angle $\delta\Omega$ in which the amplified beam is propagating.

If $\delta\Omega$ is small, as it turns out to be for most geometries, then the spontaneous term is never small compared with the stimulated term.

For example, with a brightness temperature of 10^{13} K at a wavelength of 18 cm, the ratio is

$$n_{ij}/10^3 \pi \, \delta\Omega \, \delta n_{ij}.$$

$\delta\Omega$ might be $\pi \times 10^{-10}$ sterad, when the ratio would be

$$n_{ij}/10^4 \, \delta n_{ij}$$

and so probably of the order of 1 (see Cook, 1968).

In some circumstances, it is the stimulated term in the rate equations which is negligible; the populations are then independent of the intensity of the stimulated emission and are determined by the rates at which they increase and decay in consequence of other processes. Thus the number densities are also independent of the intensity in the equation of transfer; the amplification is then said to be *unsaturated*. If, on the other hand, the number densities are not independent of the intensity, the maser is said to be *saturated*.

4.2. Unsaturated linear masers

The equation of transfer for an unsaturated amplifier, in which properties depend on just one co-ordinate, x, is

$$\pm \frac{\partial I}{\partial x} + \frac{1}{c}\frac{\partial I}{\partial t} = K_1 I + K_2,$$

where $K_1 = b_{ij} n_{ij}$ and $K_2 = a_{ij} n_i = h\nu A_{ij} n_{i}/4\pi$.

K_1 and K_2 are independent of I, and the positive or negative sign is taken according to whether the radiation is directed in the direction of x increasing or decreasing. The solutions for the two directions are independent; only for an unsaturated maser is that possible.

While radiation from many masers varies with time, there are some masers the output of which seems to be effectively constant. The simplest equation of transfer to consider is thus that for which the intensity and the populations, and therefore K_1 and K_2, are independent of time.

Suppose also that the populations do not depend on location within the amplifying gas. Such an assumption may seem gratuitously simplified; the justification for making it, and other simplifying assumptions, is that the simplest schemes should first be explored and more complex models

Amplification by stimulated emission

should be constructed only when required by the clear failure of the simplest models to account for observations.

Consider the equation

$$\frac{dI}{dx} = K_1 I + K_2,$$

where K_1 and K_2 are independent of time and position.

If radiation of intensity I_0 is incident on the amplifying cloud at $x = 0$,

$$I = (I_0 + K_2/K_1) \exp[K_1 x] - K_2/K_1.$$

It is probable that no galactic maser is unsaturated; the properties of the simple unsaturated maser serve none the less to call attention, by contrast, to those of saturated masers.

In the first place, the populations of the levels are unaffected by the radiation. Thus, in the four-level system of hydroxyl, radiation in any one transition is independent of that in all the others, since coupling between them can only occur through the influence each exerts on the populations. In the second place, the exponential dependence of I on K_1 means that relatively small differences between values of K_1, for example between transitions in the Λ-doublet quartet, or between transitions of opposite circular polarization, will lead to large differences between the corresponding intensities.

The exponential dependence of K_1 also entails some narrowing of the line profile. If $\phi(\nu)$ is the profile function for stimulated emission (determined essentially by Doppler broadening) then K_1 is proportional to $\phi(\nu)$ and the intensities at two frequencies ν_1 and ν_2 will be in the ratio

$$\exp[\phi(\nu_1) - \phi(\nu_2)] K_1 x,$$

where now K_1 is the value of K_1 at the centre frequency for which $\phi(\nu) = 1$.

Call $K_1 x$ the gain of the amplifier G.

Then if the intensity at ν is half that at the maximum,

$$\exp G(1 - \phi(\nu)) = 2,$$

or

$$\phi(\nu) = 1 - G^{-1} \ln 2.$$

Suppose the profile $\phi(\nu)$ is the Gaussian profile

$$\phi(\nu) = \exp[-(\nu - \nu_0)^2 / 2\sigma^2],$$

4.3 The saturated linear two-level maser

where σ is the half-width. Then

$$\frac{-(\nu-\nu_0)^2}{2\sigma^2} = \ln[1-G^{-1}\ln 2] = -G^{-1}\ln 2 \quad \text{if } G \gg 1$$

and

$$\nu = \nu_0 \pm \sigma[2G^{-1}\ln 2]^{\frac{1}{2}}$$

instead of

$$\nu = \nu_0 \pm \sigma[2\ln 2]^{\frac{1}{2}}$$

for the frequencies at half intensity of the Gaussian profile.

The profile of laser emission is thus narrower by a factor $G^{-\frac{1}{2}}$. Suppose for example that I_0 is 10^4 K (as from an H-II region) and that the observed intensity is 10^{13} K. The gain, the logarithm of the ratios of the emergent and incident intensities, is then about 21 and $G^{-\frac{1}{2}}$ is $1/4.6$. Thus amplification in an unsaturated maser leads to lines narrower than the spontaneous emission, but not so very much narrower, as in laboratory lasers with resonant systems.

Now suppose that the population densities, and therefore K_1 and K_2, are known functions of time, so that I is also a function of time. Then

$$\frac{\partial I}{\partial x} + \frac{1}{c}\frac{\partial I}{\partial t} = K_1 I + K_2.$$

If K_2 is supposed negligible

$$I = g(x-ct) \exp\left(\int_0^t cK_1 \, dt\right),$$

where g is an arbitrary function representing radiation travelling through the amplifying column at the speed of light, the form of g being determined by the initial conditions.

4.3. The saturated linear two-level maser

The next simplest system to consider is a saturated system in which properties again depend on one co-ordinate x. A one-dimensional maser is sometimes called a tube-maser, but there is a distinction between a strictly one-dimensional maser and a tube maser. In the tube maser it is supposed that radiation is confined by a cylindrical boundary; the transport of radiation is along the axis of the cylinder but $\delta\Omega$, the angle of emergence of radiation from the maser, varies with the distance of the emitting region from the end of the tube. Let the diameter of the tube be d, the length l and let the emitting region be at x from the input end. The

Amplification by stimulated emission

angle $\delta\Omega$ is that subtended by the exit end at the emitting region, namely $(d/(l-x))^2$ (Goldreich and Keeley, 1972, and appendix 2). In the type of maser considered here it is supposed that the direction into which a volume of gas can radiate by stimulated emission is controlled by the alignment of gas velocities, as discussed in chapter 3. Thus the solid angle into which emission occurs is not controlled by a geometrical factor but may be taken to be independent of the distance x.

The time independent equation of transfer of radiation is thus now

$$\frac{dI}{dx} = b\,\delta nI + an_i,$$

the suffixes having been dropped.

This is the equation for radiation travelling in the direction of increasing x. If radiation travels in the opposite direction

$$-\frac{dI}{dx} = b\,\delta nI + an_j.$$

In the unsaturated maser, beams travelling in opposite directions are uncoupled (because the populations do not depend on the intensity) and may be treated independently. The corresponding equations for a saturated maser are coupled through the populations and must be solved simultaneously.

If the system does not vary with time, then the rates of change of the populations are zero, namely

$$\dot{n}_i = \dot{n}_j = 0.$$

The rate equations then become

$$-uB_{ij}\,\delta n_{ij} - A_{ij}n_i - \gamma_i n_i + p + p'n_j = 0$$
$$uB_{ji}\,\delta n_{ij} + A_{ij}n_i - \gamma_j n_j - p'n_j = 0.$$

The pumping rates p and p' must also be independent of time. It will be sufficient to suppose that the rates of destruction, γ_i and γ_j, are equal to γ say; as before

$$\delta n_{ij} = (g_j n_i - g_i n_j)$$

while $u = [I^+ + I^-]/c$.

Write uB_{ij} as B. $\delta\Omega$ is supposed to be constant, as already discussed.

4.3 The saturated linear two-level maser

Let g_i and g_j be supposed equal for simplicity and let the suffixes on A_{ij} be dropped, so that the rate equations read

$$B(n_i - n_j) + (A + \gamma)n_i - p'n_j = p,$$
$$B(n_i - n_j) + An_i - (\gamma + p')n_j = 0.$$

The solutions are

$$n_i = \frac{(B + \gamma + p')p}{\gamma[2B + A + p' + \gamma]}$$

and

$$n_j = \frac{(B + A)p}{\gamma[2B + A + p' + \gamma]},$$

so that

$$n_i - n_j = \frac{(\gamma + p' - A)p}{\gamma[2B + A + p' + \gamma]}.$$

Evidently, it is possible to have inverted populations ($n_i > n_j$) with no internal pumping ($p' = 0$) provided the rate of destruction, γ, is greater than the rate of spontaneous radiation, A.

If there is no external pumping ($p = 0$), inverted populations can occur only if γ, the rate of destruction of hydroxyl, is also zero; in that case

$$n_i - n_j = \frac{p' - A}{2B + A + p'}.$$

In each case therefore, $n_i - n_j$ may be written as

$$\frac{\alpha}{\beta I + d}$$

where $I = I^+ + I^-$.

If there is external pumping and γ is not zero in the two level system,

$$\alpha = (\gamma + p' - A)p$$
$$\beta = 2\gamma B_{ij}/c$$

and

$$d = \gamma(A + \gamma).$$

If there is internal pumping and γ is zero,

$$\alpha = p' - A$$
$$\beta = 2B_{ij}/c$$

and
$$d = A.$$
n_i may similarly be written as
$$\frac{fI+g}{\beta I+d}.$$

If there is external pumping and γ is not zero,
$$f = p\beta/2\gamma$$
and
$$g = (\gamma + p')p,$$
while if there is internal pumping and γ is zero,
$$f = \beta/2$$
and
$$g = p'.$$

The number densities in a linear maser may therefore be written in the form
$$n_i - n_j = \frac{\alpha}{\beta I + d},$$
$$n_i = \frac{fI+g}{\beta I+d},$$

in circumstances of practical importance. They apply to the water maser, which is a two-level system, and they may apply to the four-level system of hydroxyl when only one transition is radiating.

The expressions for $(n_i - n_j)$ and n_i may now be substituted in the time independent equations of transfer for radiation travelling in the two directions to give
$$\left|\frac{\partial I^+}{\partial x}\right| = \frac{bI^+\alpha}{\beta I+d} + \frac{a(fI+g)}{\beta I+d}$$
and
$$\left|-\frac{\partial I^-}{\partial x}\right| = \frac{bI^-\alpha}{\beta I+d} + \frac{a(fI+g)}{\beta I+d}.$$

4.3 The saturated linear two-level maser

If radiation in one direction (take it to be of x decreasing) is suppressed, the remaining equation is

$$\frac{\partial I}{\partial x} = \frac{FI+G}{\beta I+d}$$

where I is now equal to I^+ and the superscript $+$ can be dropped.

$$F = b\alpha + af,$$
$$G = ag.$$

The implicit solution is

$$\frac{b}{F}(I-I_0) + \frac{1}{F^2}(Fd - G\beta) \ln \frac{(FI+G)}{(FI_0+G)} = x.$$

When I is close to I_0 the solution gives a nearly exponential dependence of I on x, as in the unsaturated maser, but when I is large, the logarithmic term is of less importance and I depends nearly linearly on x:

$$I = I_0 + Fx/\beta.$$

The saturated maser thus amplifies far less strongly than the unsaturated maser; at the same time, the width of the line is not decreased, because the intensity is proportional to F, and so is proportional to the line profile function $\phi(\nu)$, instead of being exponentially dependent on it. Radiation at all frequencies is thus amplified by the same factor.

Consider now what happens if the radiation in one direction is not suppressed, and suppose that the maser is asymmetrical about its centre ($x = 0$). The equations of radiative transfer for the two directions are now coupled through the factors fI and $(\beta I + d)$ which include both the intensities I^+ and I^-.

Write then

$$I^+ + I^- = I,$$

as before, and

$$I^+ - I^- = j.$$

On adding the two equations of transfer

$$\frac{\partial j}{\partial x} = \frac{(\alpha b + 2af)I + 2G}{\beta I + d}$$

$$= \frac{FI + 2G}{\beta I + d}, \text{ say,}$$

where now $F = \alpha b + 2\alpha f$; on subtracting the two equations,

$$\frac{\partial I}{\partial x} = \frac{\alpha b j}{\beta I + d} = \frac{Dj}{\beta I + d}, \quad \text{where } D = \alpha b.$$

The symmetrical boundary conditions at the two ends, $x = +L, -L$, will be

$$I^+(L) = I^-(-L),$$

$$I^+(-L) = I^-(L) = I_0, \quad \text{the incident intensity;}$$

while at the centre, $x = 0$,

$$I^+(0) = I^-(0).$$

Put $\beta I + d = Z^{\frac{1}{2}}$ so that $I = (1/\beta)(Z^{\frac{1}{2}} - d)$. Then

$$\frac{dj}{dx} = \frac{FI + 2G}{\beta I + d} = \xi + \eta Z^{-\frac{1}{2}}$$

where $\xi = F/\beta$ and $\eta = -d/\beta + 2G$. Also

$$\frac{Dj}{\beta I + d} = \frac{dI}{dx} = \frac{Z^{-\frac{1}{2}}}{2\beta} \frac{dZ}{dx}$$

or

$$j = \frac{1}{2\beta D} \frac{dZ}{dx},$$

giving

$$\frac{dj}{dx} = \frac{1}{2\beta D} \frac{d^2 Z}{dx^2}.$$

Thus, finally Z satisfies the equation

$$Z'' = \lambda + \mu Z^{-\frac{1}{2}} = 0$$

where Z'' is $d^2 Z/dx^2$,

$$\lambda = 2DC = 2\alpha b(\alpha b + \alpha f)$$

and

$$\mu = 2\alpha \beta b(d/\beta - G).$$

A general solution to the equation for Z is currently being studied by J. Viney.

Near the positive ends of the maser, I^+ is approximately proportional to distance. This may be seen directly from the fact that where I^+ is much

4.4 Other two-level systems in the steady state

greater than I^-, I and j are both nearly I^+, and dI/dx, which is the same as dI^+/dx, is then constant. Similarly, at the other end of the maser, dI^-/dx will be nearly constant. Thus at each end of the maser the radiation travelling outwards is almost proportional to distance along the maser. The argument is however unable to answer the question how the respective intensities depend on the incident radiation at the opposite end, and the answer to that depends on the solutions in the region where I^+ and I^- are comparable. Unpublished numerical solutions by M. S. Middleton show that the emergent intensities are very insensitive to the incident intensities over a wide range of the latter and are very nearly equal at the end points, as well as at the centre of the maser.

Since the differential equations for I^+ and I^- are, when I^+ or I^- is large,

$$\frac{\partial I^{\pm}}{\partial x} = \pm \frac{b\alpha + af}{\beta}$$

it follows that the solutions for I^+ and I^- approach respectively

$$I^{\pm} = \pm \left(\frac{b\alpha + af}{\beta}\right)x + \text{const.}$$

The factor $(b\alpha + af)/\beta$ is, when spontaneous emission is small compared with stimulated emission, proportional to an effective pumping rate, p_{eff}, say. When, for example, the pumping occurs at a rate pn_i from the lower level to the upper level, p_{eff} is $(p - A)$. The factor is also proportional to $h\nu/\Delta\nu$. Thus in the linear region where I is large, all masers behave in the same way, scaled by a factor of $h\nu p_{\text{eff}}/\Delta\nu$.

Now while elementary arguments determine the proportionality factor in the expression for I^+ and I^-, they do not determine the zero from which x is to be measured, or what is effectively the same thing, the value of the constant. Middleton's solutions for the two-level linear maser show that to a very close approximation, x is to be measured from the centre of the maser and the constant is then zero (figure 4.1).

4.4. Other two-level systems in the steady state

The considerations set out in the previous section apply to the simplest model of a maser, the line maser, in which the properties depend only on one co-ordinate and the solid angle into which radiation is emitted does not vary with distance along the maser. Reasons have been given for thinking that such a model, while no doubt oversimplified, none the less reproduces important features of actual masers; at the same time it must be recognized that other geometries are plausible. It was seen in chapter 3

Amplification by stimulated emission

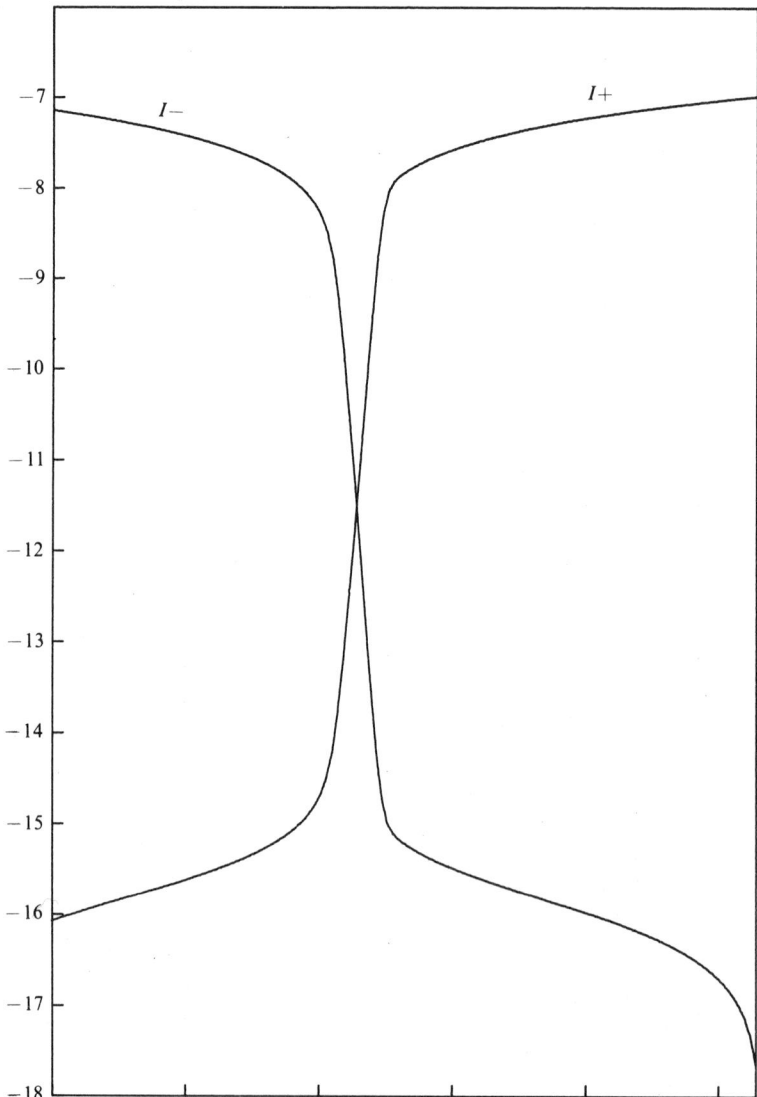

Figure 4.1. Intensity of radiation in a two-level linear maser. Numerical solutions by M. S. Middleton. Abscissa: distance along the maser. Ordinate: intensity of radiation on logarithmic scale.

that hydroxyl and water sources appear to lie in spherical shells surrounding H-II regions (thin shells) or infra-red stars (thick shells) and so it is reasonable to seek solutions of the equations of transfer in spherical shells.

4.4 Other two-level systems in the steady state

In spherical geometry, radiation may be travelling in all directions. Let θ be the angle between the direction of the radiation and the radius vector drawn from the centre of the spherical shell. Denote $\cos\theta$ by μ.

If the system is supposed to be spherically symmetrical the intensity is a function of μ and radius only; the radiation density is

$$u = \frac{1}{c}\int_{-1}^{+1} I(\mu, r)\, d\mu.$$

Let s be the line element in the direction of the travel of the radiation. Then

$$\frac{\partial I}{\partial s} = \frac{1}{\mu}\frac{\partial I}{\partial r} + \frac{1-\mu^2}{r}\frac{\partial I}{\partial \mu}$$

and the equation of transfer for a transition between levels i and j reads

$$\frac{1}{\mu}\frac{\partial I_{ij}}{\partial r} + \frac{1-\mu^2}{r}\frac{\partial I_{ij}}{\partial \mu} = I_{ij}B_{ij}\delta n_{ij} + A_{ij}n_i.$$

If the system is a two-level maser or if amplification occurs in only one of a group of transitions, then δn_{ij} is, as before, a function of u_{ij} of the form $a/(\lambda u_{ij}+d)$ while n_{ij} is of the form $(ku_{ij}+l)/(\lambda u_{ij}+d)$.

Analytical solutions are not known for spherical shells, although some numerical work has been done (Litvak, 1970). Some idea of the results can be obtained simply by considering how the length of the path of radiation through a shell varies with direction. The amplification is proportional to the exponential of the optical thickness for an unsaturated maser, or to the thickness itself for a saturated maser, and so paths with the greatest length through the shell should amplify most strongly.

Suppose then that the molecules in the shell amplify the radiation emerging from hot gas inside the shell, so that the relevant distance is that from the inner surface to the outer. Clearly this distance is greatest when the radiation is travelling tangentially to the inner boundary of the shell, and so it might be expected that radiation amplified by a spherical shell of gas in this way would appear as a bright rim at the edge of the cloud of gas. The elementary source spots do indeed appear to lie around the rims of H-II regions, but of course do not form a continuous bright rim, for reasons which have already been suggested.

No doubt, if it were thought to be justified, the equation of transfer could be solved numerically for any particular geometrical condition. In any case, if only one transition is amplifying, then the number densities depend on the density of the amplified radiation in the same way as for the two-level maser. The difficulty in the way of solutions of systems other than the line maser is that the radiation density is not just the sum of densities in two directions but must be found by integration over all angles.

Goldreich and Keeley (1972) (see also Lang and Bender, 1973) have solved the equations for steady state maser action between two levels in a spherical cloud of gas and in a cylinder when the source is the spontaneous emission. They express their results for the spherical maser in terms of a coefficient α, equal to $(P/\Delta P)(A/\Gamma)$ where P is the mean rate at which molecules enter the two levels, ΔP is the difference between the rates, Γ is the decay constant from the levels and A is the Einstein coefficient of spontaneous emission.

The spontaneous radiation from the gas in the sphere is supposed to have a brightness temperature T_i. Goldreich and Keeley distinguish three regions. In the first, the maser is unsaturated throughout the sphere; the sphere must have a dimensionless radius less than R_1 where R_1 is given by

$$\exp[2R_1] = 4R_1/\alpha.$$

The apparent size of the maser as seen from a great distance is then $R^{\frac{1}{2}}$ and the brightness temperature of the amplified radiation is $e^{2R}T_i$, just as for an unsaturated linear maser.

In the second condition, there is a central core in which the maser is unsaturated, having a radius a given by

$$a^3 e^{-2a} = \alpha R^4/\gamma.$$

The apparent size is then $a^{\frac{1}{2}}$ (and so relatively smaller than the unsaturated maser although greater absolutely) and the brightness temperature is $(4R^3/3\alpha a)T_i$.

Thirdly, the maser may be saturated throughout with a central core of radius a in which the intensity is constant and an outer shell in which the intensity increases linearly outwards. The radius of the core is given by

$$a = 1.3\, R\alpha^{\frac{1}{4}},$$

the apparent size of the maser is $0.7a$ and the brightness temperature is $(5.2\, R/\alpha^{\frac{3}{4}})T_i$.

Goldreich and Keeley distinguish in a similar way, unsaturated, partially saturated and saturated conditions in a cylindrical maser, expressing their results in terms of a scale length L equal to

$$\frac{4\pi}{Bh} \cdot \frac{\Delta \nu}{\nu} \cdot \frac{1}{\Delta N},$$

where B is the Einstein coefficient of stimulated emission, $\Delta \nu$ is the line

4.4 Other two-level systems in the steady state

width, ν is the frequency and ΔN is the relative difference of populations in unsaturated conditions (this is equivalent to the scale factor discussed earlier).

Goldreich and his collaborators have also discussed the effect of transitions between maser levels that are the consequence of absorption and re-emission of infra-red emission by the amplifying gas. Goldreich, Keeley and Kwan (1973b) point out that if a cloud of hydroxyl is dense enough to exhibit maser amplification, it will also be very thick optically for transitions between rotational levels and in consequence there will be very rapid relaxation between the Λ-doublet levels. They show that if the rate of relaxation exceeds the rate of decay from the levels, the maser radiation will be emitted in a narrow solid angle and the excess population will be saturated in all the magnetic sub-levels. The polarization of the radiation will thus be affected by the cross relaxation. In standard maser theory it is commonly said that the width of the radiated line is reduced in an unsaturated maser (section 4) but increases again in a saturated maser. Goldreich and Kwan (1974) show that in the presence of cross-relaxation by trapped infra-red radiation, the width of the line remains narrow provided

$$A \gg R(e^{h\nu/kT} - 1)$$

where A is the rate of spontaneous emission, R the rate of stimulated emission and ν the frequency.

A feature common to many models of maser amplifiers is that the central portion of the maser is unsaturated. Three models in particular have been studied. The results in section 4.3 are for a line maser amplifying radiation from an external source, intensities incident on the two ends being in general different. The numerical results obtained by Middleton (referred to in section 4.3) show that over a wide range of ratios of the incident intensities there is a small unsaturated region close to the centre of the maser. Goldreich and Keeley's (1972) analysis of a maser amplifying spontaneous emission depends on matching solutions for saturated outer portions, to that for an unsaturated inner section. It should be noted, by the way, that whereas in the asymptotic solutions for the line maser, the intensity is proportional to length, in that for the tube maser it is proportional to the cube of the length, the difference being due to the fact that the solid angle into which radiation is emitted is constant for the line maser, but varies with distance along the tube maser. The solutions also differ from those for the line maser given above because they deal with amplified spontaneous emission instead of amplification of some specified incident radiation.

Amplification by stimulated emission

Yet again, the solutions of Goldreich and Keeley (1972) and Lang and Bender (1973) for the spherical maser have unsaturated central cores, which play an essential part in reducing the apparent size of the maser to something very much less than the size of the sphere of amplifying gas. Of course the fact that rays passing through the centre are longer than any chordal rays means that the gain along the diameter is greater than the gain along any chord, but more important is the fact that a diametral ray passes through the central unsaturated region of high gain and the effective size of the maser as seen from a distance is determined by the spread of rays which pass through the unsaturated core.

Eveything that has been said so far applies to two-level masers in which the pumping rates are independent of, or proportional to, the numbers in the lower of the two levels, and an argument has been given for supposing that the restriction on the pumping rates may well be satisfied in some actual circumstances (section 4.1). The two-level model of this type may well be a suitable model for the water maser where the pumping rate may be independent of the numbers in the two levels of the maser. It may also apply to the hydroxyl maser, although that may seem unlikely at first sight.

In the first place, intensities of infra-red transitions involving the levels participating in the maser and which determine the numbers in those levels are not in principle independent of the intensity of maser radiation so that the pumping rates cannot be expressed as quantities independent of, or proportional to, the numbers in the lower maser levels. Yet in actual circumstances, the geometrical distribution of the amplifying columns or spheres of hydroxyl gas may lead, as argued in section 4.1, to a field of infra-red radiation locally independent of a particular maser; in that case the restriction on pumping rates would be satisfied.

In the second place it is observed quite generally and possibly without exception (Cook, 1975, and chapter 3) that from any one direction in the maser, radiation is emitted from one only of the four possible ground state transitions in OH (there is insufficient evidence to say if the same is true for transitions in rotationally excited states, but in general only one or two of the possible transitions of excited states are ever detected). Accordingly it is proposed that some filter mechanism suppresses radiation from all but one of the transitions or alternatively that the pumping process leads to a pattern of inversion such that amplification can occur in only one transition. In those circumstances, there is only one set of equations of radiation transfer to be considered – those for the transition that is radiating – and in the equations for the populations there is only one set of stimulated terms – again those proportional to the density of

4.5 Polarization

the energy of the one radiating transition. With such a simplification it is easy to show that the equations of transfer have the same form as those for the elementary two-level maser and all the results for tube line and spherical masers will apply to hydroxyl radiation from a single transition.

4.5. Polarization

From one point of view, the problem of polarization is a particular case of a multi-level maser in which radiation from certain transitions is suppressed. Radiation is circularly polarized because transitions occur preferentially between those magnetic sub-levels for which the magnetic quantum numbers differ by $+1$ or -1 and if by some process one type of transition is allowed while the other is suppressed, circular polarization of one sense will be amplified. Similarly, if only those transitions with $\Delta m_F = 0$ were to occur, linearly polarized radiation would be amplified, while if all transitions could occur and there were no distinction between magnetic sub-levels, the radiation could be considered to come from a single unpolarized transition. The problem to be tackled in the theory of amplification of polarized radiation is how far transitions with different Δm_F can be considered separately and how far one may influence the other, and the answers will depend on the ratio of the widths of the energy levels to the Zeeman shift, and the effect of Faraday rotation if there are free electrons in the amplifying gas.

The suppression of one type of transition and the enhancement of another may be inherent in maser amplification itself on account of the non-linearity of the physical processes – of two competing transitions, the gain is greater in the one with greater intensity of radiation. Thus it may be possible for one sense of polarization by chance initially slightly more intense, to come to dominate the other. Such a process is well known in laboratory masers and the quantum mechanical theory for a celestial maser has been given by Heer (1966) and Heer and Settles (1967*a,b*). The work of Bromley (1971) is also relevant. However, the theory of Heer and Settles assumes that the radiation being amplified is highly monochromatic (as is usually the case in laboratory studies) but Bender (1967) has pointed out that the radiation being amplified by celestial masers covers a broad band and has shown that polarization does not then develop. In the laboratory work to which the theory of competitive amplification is applicable, the wavelength and spread of the radiation are determined by the incident field, independently of the amplification process, whereas in celestial masers it is the amplification process through its dependence on relative velocities of molecules that determines the mean wavelength and profile of the amplified radiation.

Amplification by stimulated emission

If theories involving competition between transitions do not seem to be able to explain polarization, then one must turn to Zeeman splitting by a magnetic field. A small Zeeman splitting may indeed be involved in competitive theories to produce the initial slight imbalance between transitions, but it must not be too great or again the theory fails. In other theories, on the other hand, the Zeeman shift, differing for different senses of polarization, plays an essential part in suppressing one and enhancing the other. The mechanism suggested in chapter 3 is of this type – the Zeeman shift cancels a Doppler shift for one sense of polarization but reinforces it for the other.

Goldreich, Keeley and Kwan (1973a) have developed the theory of polarization to include the effects of relaxation between sub-levels, the consequence in particular of collisional transitions which tend to equalize the intensities of transitions. They derive the equations of transfer for polarized radiation from the equation of motion for the matrix of the densities of states of the molecules in the presence of a magnetic field.

They take Γ to be the decay constant of the molecules in excited levels, R the rate of stimulated emission, $g\Omega$ the Zeeman splitting[†] and $\Delta\omega$ the bandwidth of the amplified radiation. Then if the maser is unsaturated, R is less than Γ and Goldreich, Keeley and Kwan show that stimulated radiation will be unpolarized unless

$$g\Omega > \Delta\omega,$$

that is to say, unless there is no overlap of the radiation with opposite senses of polarization. On the other hand, in a saturated maser R is greater than Γ and radiation will be unpolarized unless

$$g\Omega > (R\Gamma)^{\frac{1}{2}}.$$

If in addition $g\Omega$ is greater than $\Delta\omega$ the radiation will be 100 per cent circularly polarized.

The theory of Goldreich, Keeley and Kwan predicts that radiation of both senses of polarization should be emitted and does not indicate how one might be suppressed – what it does predict is the minimum value of the magnetic field necessary for polarization to be possible. Thus it is still necessary to find a mechanism that will suppress one sense of polarization, such as the one described in chapter 3; for any such mechanisms to operate it is necessary, as was supposed in the discussion in chapter 3, that the Zeeman shift ($g\Omega$) exceeds the Doppler width of the radiation ($\Delta\omega$).

Goldreich, Keeley and Kwan point out that whereas the splitting factor g is about 1 for hydroxyl (chapter 2) it is only about 8×10^{-3} for water and

[†] g is the splitting factor and Ω the electron gyro-frequency in the applied field.

4.6. Masers varying in time

so the Zeeman splitting will be less than the bandwidth (about 100 kHz) for fields less than 4×10^{-3} T. That no doubt accounts for the absence of circular polarization in radiation from water.

4.6. Masers varying in time

It was shown in section 4.1 that the equation of transfer for a line maser varying in time is

$$\frac{\partial I}{\partial x} + \frac{1}{c}\frac{\partial I}{\partial t} = Ib_{ij}\,\delta n_{ij} + a_{ij}n_i$$

and that the solution for an unsaturated maser is

$$I = g(x-ct)I_0 \exp\left[x\int_0^t b_{ij}n_{ij}\,\mathrm{d}t\right]$$

if a_{ij} can be neglected.

Solutions of the equation for the unsaturated maser have a characteristic time of the order L/Gc where L is the length of the amplifying column and G is the (logarithmic) gain of the amplifier.

They represent a wave of arbitrary form, $g(x-ct)$, passing through the medium at the speed of light (the refractive index was assumed to be that of vacuum in setting up the equation of transfer) and amplified by the exponential factor.

These are the only analytical solutions for a time-varying maser that are known. None has been obtained for an unsaturated maser of any other geometry and none is known for any saturated maser.

Furthermore, until recently, no numerical solutions had been obtained, the lack of which was felt in discussing the mode-competition theory of polarization as well as the interpretation of any maser observed to vary in time. Now however numerical methods have been developed by M. Salem and also by M. S. Middleton for the solution of the general two-level time-dependent maser. It is found that when a maser is suddenly switched on, the output rises to a peak, undergoes damped oscillations and then settles down to a steady value; the period of the oscillations is about equal to the time for radiation to travel back and forth through the column (figure 4.2). Some solutions have also been obtained for a pumping rate varying sinusoidally with time; the maser output varies sinusoidally with the same period (figure 4.3).

The results may be relevant to the suggestion of Deguchi (1974) that maser amplifiers are unstable (the mode-competition theory of polarization is of course an example of the instability). The two-level linear maser

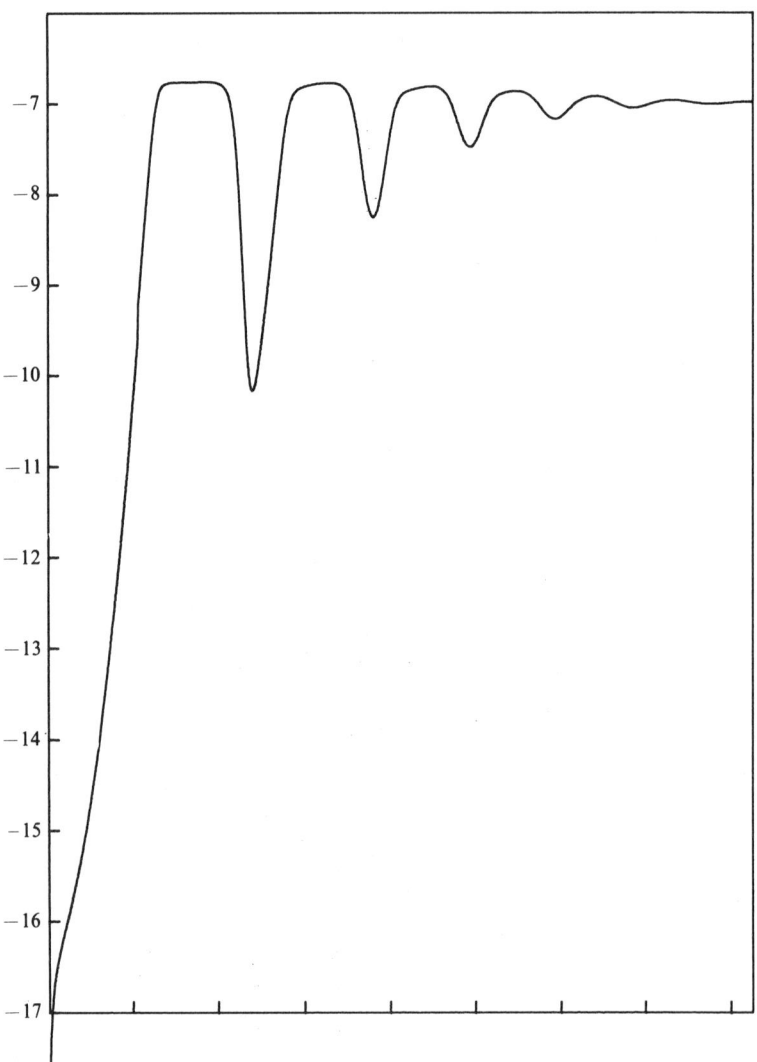

Figure 4.2. Variation of intensity from a two-level linear maser after sudden switching on. Numerical calculations by M. Salem. Abscissa: Time after switching on. Ordinate: Intensity of emitted radiation on logarithmic scale.

seems, from numerical results so far obtained, not to show instability, but more work should be done before that is clear.

The behaviour of the four-level maser in time has not so far been explored. If the pumping rate is again independent of the numbers in the lower levels, or proportional to them, then the equations again reduce to

4.6 Masers varying in time

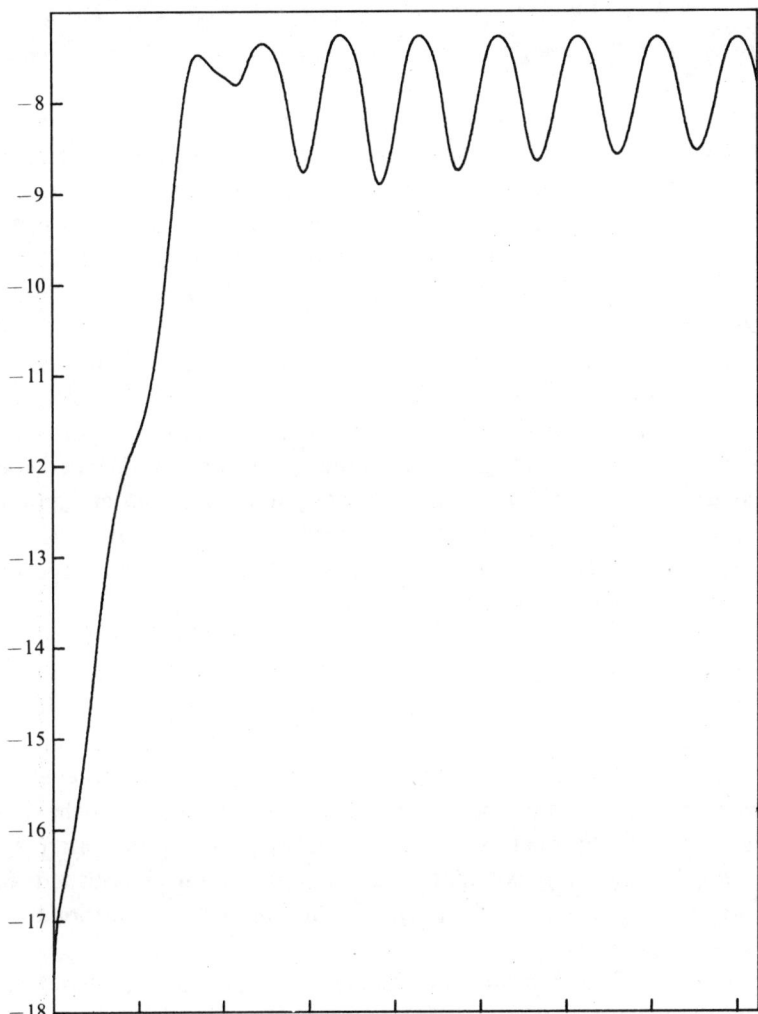

Figure 4.3. Radiation from a two-level linear maser with sinusoidally varying pump rate. Numerical calculations by M. Salem. Abscissa: Time. Ordinate: Intensity of emitted radiation on logarithmic scale.

those for a two-level system, but of greater complexity than for an elementary two-level maser. Nothing as yet is known of their solutions. An important problem for the future is to see if a multi-level maser is unstable – does competition between transitions lead to one dominating over the other in the course of time?

Amplification by stimulated emission

4.7. Summary

The aim of this chapter has been to give a phenomenological account of amplification by stimulated emission and to eschew quantum mechanical considerations. A fully satisfactory description would no doubt use, as do descriptions of laboratory lasers and masers, the formalism of the density matrix, but it may be thought that in the present state of knowledge of conditions in maser sources, a shallower approach is all that is justified. Even so, it is evident that exact solutions can be given only for the simplest models, and those probably not very representative of actual conditions.

The major gap in understanding is that we do not assuredly know how to account for the small angular sizes of sources, the isolated distribution of individual sources, the single sense of polarization in hydroxyl and, again in hydroxyl, radiation in only one transition. Until there is better physical understanding of these features, an acceptable detailed theory cannot be produced. At the same time, the way in which very high intensities arise is understood in general terms, as are some properties of linear masers and masers with spherical symmetry.

Possibly the greatest lack is the absence, so far, of any but the sketchiest account of lasers varying in time. In view of the rapid variations that are observed, especially in water masers, the development of a theory of some elementary masers that vary in time is urgent. Among other problems which require study are the behaviour of a multi-level maser when radiation is amplified in only one transition, and the behaviour in time of a multi-level system for if, as is possible, the latter shows instability involving switching from one transition to the other, it could form a model to account for the occurrence of a single transition from an isolated direction in space.

Beyond all the gaps in a phenomenological approach to maser theory stands the fact that very little has been done to produce a quantum theory of amplification using a density matrix formulation. Some attempts have been made, but attention has been concentrated on the phenomenological approach adopted here, and even though the physical conditions in celestial masers are poorly understood, more work on the density matrix formulation might illuminate some general properties.

Emphasis has been laid in this summary on gaps in our theoretical understanding, as is justified, yet while we are still far from the goal of applying theory in detail to interpret particular masers, we can draw some general results from what theory has already been done. In the first place, saturated two-level linear masers fit very well the expression

$$I^{\pm} = \pm kx$$

4.7 Summary

where k is proportional to $p_{\text{eff}}\nu/\Delta\nu$ and x is measured from the centre of the amplifying column.

That expression is almost independent of the intensity of the incident radiation that is amplified.

These results, the linear form, the scaling factor and the independence of input intensity, are so simple that it is reasonable to conjecture that they will apply to more complex geometries and the multi-level systems, although in the latter case, the definition of an effective pumping rate may not turn out to be so simple as for the two-level maser.

The studies of Goldreich and his colleagues have clarified the conditions in which polarization may occur, although not the mechanism which selects a single sense, and they have also shown the importance of cross relaxation in the presence of an intense thermal radiation field. These results again may well generalize.

Finally, the small apparent sizes of masers may arise in one of two ways – either as a consequence of alignment of gas velocities to produce a linear maser of small cross-section or as a consequence of the unsaturated core at the centre of a spherical maser.

5

Pumping schemes

5.1. Introduction

Many schemes have been suggested for inverting the populations of hydroxyl and water, and it is convenient to collect them together before coming to the interpretation of observations of stimulated emission.

In water, populations of a single pair of levels ($6_{16}, 5_{23}$) have to be inverted, but hydroxyl populations may be inverted among one or more of four pairs of levels. Let the levels in the rotational ground state of hydroxyl be labelled in descending order 4, 3, 2, 1 as in chapter 2. Transitions for which $\Delta F = 0$, namely $4 \rightarrow 2$ (1667 MHz) and $3 \rightarrow 1$ (1665 MHz), have the greatest probabilities and are known as *main line transitions*; if the Λ-doublet populations as a whole were inverted and there were no distinction between levels 4 and 3 and 2 and 1, that is

$$n_4/g_4 = n_3/g_3, \quad n_2/g_2 = n_1/g_1$$
$$\text{and} \quad n_4/g_4 + n_3/g_3 > n_2/g_2 + n_1/g_1,$$

then all transitions could develop maser action but the main lines having the greater strengths, would be favoured. On the other hand, if the Λ-doublet as a whole were not inverted so that

$$n_4/g_4 + n_3/g_3 = n_2/g_2 + n_1/g_1,$$

but there were inequalities between levels with different values of F, there would be no inversions of population for the main line transitions and stimulated emission could occur only in the weak or *satellite* lines. Thus, if

$$n_4/g_4 > n_3/g_3, \quad n_2/g_2 > n_1/g_1$$
$$\text{and} \quad n_4/g_4 + n_3/g_3 = n_2/g_2 + n_1/g_1,$$

n_4/g_4 would be greater than n_1/g_1 and there could be stimulated emission at 1720 MHz, whereas if

$$n_4/g_4 < n_3/g_3, \quad n_2/g_2 < n_1/g_1$$
$$\text{and} \quad n_4/g_4 + n_3/g_3 = n_2/g_2 + n_1/g_1$$

n_3/g_3 would be greater than n_2/g_2 and there could be stimulated emission at 1612 MHz. Accordingly, it is possible to group postulated pumping

5.1 Introduction

schemes for hydroxyl according to whether they distinguish between hyperfine levels of the Λ-doublet or whether they invert the Λ-doublet as a whole, the former yielding maser radiation in the satellite lines and the latter giving maser radiation predominantly in the main lines. This is the distinction between the sources in classes II and I of Turner (1970) and it seems therefore to be a significant way of classifying pumping processes.

In some classes of hydroxyl source, especially those of Turner's type II, only one transition occurs, usually that at 1612 MHz, and it may then be that it is the pumping process that determines which transition radiates. But in other sources, particularly of type I, all transitions may radiate, although, as has been seen in chapter 3, only one does so from any one discrete direction. In those sources it seems that the Λ-doublet is inverted as a whole and that the transition that radiates along a particular line of sight is picked out by a filter mechanism as suggested in chapter 3.

Another way of classifying pumping processes is according to whether they involve photon processes, or collisions without chemical change, or a chemical action. In photon processes it is supposed that molecules absorb radiation and make transitions to higher states, from which they decay to levels of the Λ-doublet, one or more of the latter being preferred because of differences in rates of decay from the higher states. In this respect it should be noted that some transitions may be forbidden by selection rules based on angular momentum or parity. If radiation processes are to determine populations in this way, collisions must not be frequent enough to thermalize populations. A difficulty with photon processes is that to each 18 cm photon radiated there must correspond at least one pumping photon of higher energy but the total numbers of 18 cm photons are very great (chapter 3) and in general it seems unlikely that sufficient pumping photons are available from known sources associated with hydroxyl masers.

Collisions, especially with charged particles, may be expected to excite molecules, either inverting populations directly or indirectly through radiative decays from higher levels. Electrons and protons with energies of up to 10 eV may be found in the ionization front surrounding H-II regions and it has been suggested that they may be responsible for pumping hydroxyl; unfortunately the characteristics of such collisions are not known in detail. Collisions may also bring about chemical reactions, in particular, collisions of various particles on water may form hydroxyl in excited states.

If populations of levels are inverted, as they must be in a maser, then the gas is not in thermodynamic equilibrium, and further, if it is bathed in a radiation field which produces the inversion, then neither is that field thermal.

A condition for a radiation field to have a Planckian distribution and to be in equilibrium with matter in which the populations of energy levels have a Boltzmann distribution, is that the radiation should be confined by the matter long enough for statistical equilibrium to be established through detailed balancing. In the usual derivations of the Planckian distribution and the Einstein coefficients of stimulated emission and absorption, it is supposed that the radiation is confined in a cavity, so that it remains in contact with matter for an indefinite time. If inverted populations are to be produced by radiative processes, radiation must be able to escape from the maser region in a time short compared to that needed to set up a Boltzmann distribution of the energy levels in all the matter. Considering the matter to be a mixture of dust and gas, and supposing the gas to have only a few well separated energy levels in the range of frequency of interest, it follows that the density of dust must be low enough so that it does not absorb and then re-emit to thermalize a substantial proportion of the radiation that might invert the energy levels of the gas.

It might seem therefore that in a large or dense enough cloud of gas or gas and dust, inversion of populations could only occur towards the outside of the cloud, from which radiation could escape sufficiently rapidly. Going further, Jefferies (1971) has argued not just that inversions may occur in such regions but that they will occur there, irrespective of the detailed mechanism by which they are brought about. The essence of Jefferies's argument is that because the strengths of cascading transitions in a cloud of gas are in general all different, the optical depths will differ among transitions. Thus some transitions will be coupled but others decoupled if the radiation escapes from the gas (ter Haar and Pelling, 1974) and it is possible for groups of levels to be established in which the populations are coupled, such that they are at the same time all either greater, or less, than populations of levels in other groups (de Jong, 1973; Pelling 1975a,b). Anomalous absorption can also be explained on similar lines (Pelling and ter Haar, 1975).

Somewhat similar considerations apply to inversions produced by collisions. If collisions, whether of the colliding particles or of excited molecules of the gas with dust or other gas, are sufficiently frequent, the populations will be brought into thermal equilibrium. It follows that whatever the nature of the inverting process – whether by collisions with energetic particles, by radiation or by chemical processes, there is a limit to the density of matter (dust and gas) in the maser region, above which

5.2 Pumping of hydroxyl – photon processes

the process cannot occur. On the other hand, the density of maser gas must be above some lower limit or no significant amplification by stimulated emission will occur.

Most of the inverting processes that have been proposed are listed in the remainder of this chapter, with some brief account of the suggested mechanism. Detailed discussions and criticism have been deliberately omitted for two reasons. First, neither the observational data nor, often, the molecular data are sufficiently well known to justify much discussion. Secondly maser radiation is of such relatively common occurrence that it may be that some rather general mechanism operates as it does in many laboratory gas lasers. The general considerations in this section may therefore, in the present state of understanding, be more pertinent than detailed analysis of a particular scheme.

5.2. Pumping of hydroxyl – photon processes

The first pumping schemes to be suggested depended on the absorption of ultra-violet radiation, either hydrogen Lyman-alpha (Cook, 1966, no longer plausible) or near ultra-violet radiation (Litvak and others, 1966). In the latter scheme hydroxyl molecules near to a source of ultra-violet radiation would absorb that radiation and be excited into the first excited electronic state ($A: {}^2\Sigma_{\frac{1}{2}}$). Only those levels with negative parity can decay to the upper Λ-doublet levels of the ground state, because the latter have positive parity. The hydroxyl molecules nearest to the source absorb preferentially at the stronger transitions, leaving the remaining incident radiation relatively enhanced in the weaker lines. By considering in detail the probabilities of transitions, Litvak argued that the residual ultra-violet radiation would populate those levels of the first excited vibrational state that would decay preferentially to the upper levels of the Λ-doublets of the ground states; it being assumed that collisions would not be numerous enough to thermalize populations. Litvak traced out (figure 5.1) how inverted populations would be established for different transitions as the ultra-violet radiation traversed the cloud. Because transitions from a given level of negative parity in the ${}^2\Sigma_{\frac{1}{2}}$ state to each of the upper levels of positive parity in the ground state quartet are equally likely, the mechanism inverts the Λ-doublet as a whole and would lead to stimulated emission at 1667 and 1665 MHz.

Litvak (1969) has also proposed a pumping mechanism depending on the absorption of infra-red radiation by the ground state of hydroxyl, leading to transitions in the ${}^2\Pi_{\frac{3}{2}}$, $J = \frac{5}{2}$ level, at 8400 m^{-1} above the ground state, and the ${}^2\Pi_{\frac{1}{2}}$, $J = \frac{1}{2}$ at 12 600 m^{-1} above the ground state.

Pumping schemes

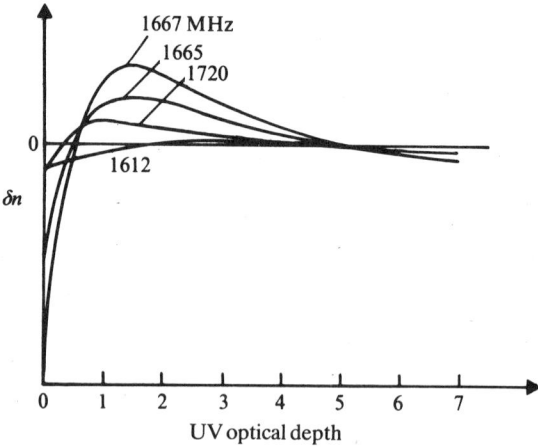

Figure 5.1. Variation of inversion of populations (δn) in the hydroxyl ground state as a function of optical depth in the ultra-violet. (ter Haar and Pelling. © 1974. The Institute of Physics.)

A diagram of some of the possible transitions is given in figure 5.2. If radiation with a continuous spectrum, in particular thermal radiation, falls on a cloud of hydroxyl molecules, the molecules will absorb preferentially at frequencies with the strongest transition probabilities; thus deeper in the cloud radiation in the weaker transitions is relatively stronger. Levels in the $^2\Pi_{\frac{3}{2}}$, $J=\frac{5}{2}$ and $^2\Pi_{\frac{1}{2}}$, $J=\frac{1}{2}$ states excited by these weaker transitions decay preferentially to one or other of the levels of the upper pair of the ground state set of levels and so deep in the cloud the

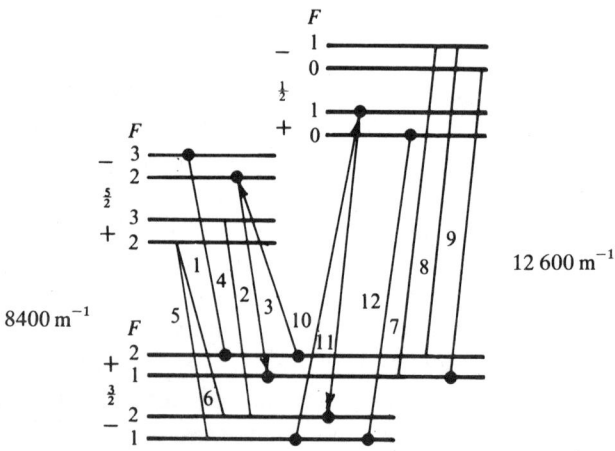

Figure 5.2. Far infra-red transitions in hydroxyl. (ter Haar and Pelling. © 1974. The Institute of Physics.)

5.2 Pumping of hydroxyl – photon processes

latter level has a high population. The mechanism thus discriminates between levels with different values of F. If radiation at 8400 m^{-1} is more intense, the $F = 1$ level would be more populated and stimulated emission could occur at 1612 MHz while if radiation is more intense at 12 600 m^{-1}, stimulated emission could occur at 1720 MHz, the $F = 2$ upper level being preferred (figure 5.3).

A third mechanism of the same type as the preceding two involves the absorption of near infra-red radiation at a wavelength of 2.8 μm which would excite higher vibrational levels; symmetry rules prevent inversion of the Λ-doublet as a whole (Litvak, 1969; Litvak and Dickinson, 1972). More recently Elitzur and others (1976b) have put forward a scheme which involves absorption of radiation at a wavelength of 35 μm, corresponding to transitions from the ground state to the $^2\Pi_{\frac{1}{2}}$, $J = \frac{5}{2}$ state. They show that if the hydroxyl gas is optically thin to the infra-red radiation, then no inversion of the Λ-doublet of the ground state will occur, whereas if the gas is optically thick, the Λ-doublet populations will be inverted so as to allow stimulated emission at 1612 or at 1720 MHz. The 1612 MHz levels are always inverted.

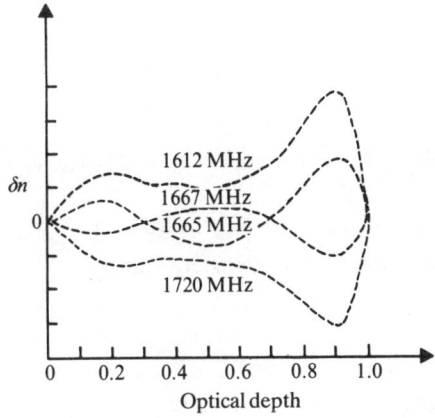

Figure 5.3. Variation of inversion of populations (δn) in the ground state of hydroxyl as a function of optical depth in the near infra-red. (ter Haar and Pelling. © 1974. The Institute of Physics.)

It will be seen that of the photon processes suggested, those that rely on absorption of ultra-violet radiation lead to emission at 1667 and 1665 MHz while those that rely on absorption of infra-red radiation lead to emission at 1720 or 1612 MHz. The reason is that the structures of levels in the $^2\Sigma_{\frac{1}{2}}$ electronic excited states, populated by absorption of

Pumping schemes

ultra-violet radiation, and those in the $^2\Pi$ states populated by absorption of infra-red radiation, are different. Turner (1970) has suggested that continuum radiation at 18 cm could invert the Λ-doublets.

5.3. Collisions on hydroxyl

Johnston (1967) proposed that collisions with electrons near the edge of an H-II region could invert populations of the Λ-doublet and predicted that linearly polarized emission would occur at 1667 and 1665 MHz. In 1970 he discussed what would happen if colliding particles were streaming in a magnetic field and proposed that circular polarization could be developed.

Somewhat related to this mechanism is that of Melrose (1973) who suggested that inversion of the Λ-doublet levels could occur if the hydroxyl were in a plasma with the plasma frequency equal to the hydroxyl transition frequency.

5.4. Formation of hydroxyl in excited states

Three proposals have been made for the formation of hydroxyl in excited states by pre-association through the combination of atomic hydrogen and oxygen.

Symonds (1965) considered the reaction

$$O^- + H \rightarrow OH^+ + e$$

which he states has a maximum cross-section for a relative velocity of 25 km s^{-1}.

Hughes (1967) suggested that the collision of hydrogen and oxygen would give rise to hydroxyl in an electronic excited state:

$$O + H \rightarrow OH \text{ in } {}^2\Sigma^+_{v=0}\dagger.$$

Solomon (1968) discussed the potential energy curves of different excited states of hydroxyl and considered that the process would be

$$H(^2S) + O(^3P) \rightarrow OH(^2\Sigma^-)$$
$$\rightarrow OH(^2\Sigma^+_{v=2})$$
$$\rightarrow OH(^2\Pi) + 306.4 \text{ nm}.$$

He pointed out the importance of rates of destruction of hydroxyl by collisions ($H + OH \rightarrow H_2O$) in relation to the rates of pre-association.

---

† The $^2\Sigma$ state is an electronic excited state in OH, $v = 0$ means that the lowest vibrational state is involved.

5.5 Formation of hydroxyl from water

5.5. Formation of hydroxyl from water

Gwinn and others (1973) have considered in detail collisions between atomic and molecular hydrogen and water that might lead to the production of hydroxyl in excited states. They discuss the reactions:

$$H + H_2O + 5.19\ eV \rightarrow OH^* + 2H \tag{1}$$

$$H + H_2O + 0.69\ eV \rightarrow OH^* + H_2 \tag{2}$$

$$H + OH \rightarrow OH^* + H \tag{3}$$

$$H_2 + OH \rightarrow OH^* + H_2. \tag{4}$$

The inversion of populations in the ground state (or in one of the higher rotational states) involves two elements, both of which are discussed by Gwinn and others (1973). The collision itself will in general lead to the production of hydroxyl in higher rotational states; Gwinn and his colleagues show that in suitable circumstances the upper of the Λ-doublet levels of those higher states will be populated preferentially. The second element involves decays from the upper rotational states to the states between which the stimulated emission is to take place; radiative transitions must dominate and the relative rates must be such that the upper Λ-doublet levels of the stimulated transitions are preferred.

The result of a collision between water and hydrogen depends on the relation between electron orbitals in the water and in the hydroxyl that is produced from it. Gwinn and his colleagues measure the degree of spatial orientation of a wave function, Ψ, by the quantum mechanical average of $(y^2 - x^2)$ where x and y are Cartesian co-ordinates in and perpendicular to the plane of rotation of an OH molecule. The average is

$$\langle \Psi | (y^2 - x^2) | \Psi \rangle$$

and some values obtained by Gwinn and his colleagues are given in table 5.1.

The spatial orientation of Λ-doublet wave functions is similar to the spatial orientation of the valency orbitals in a molecule. Gwinn and others argue that in a collision between a hydrogen atom and a water molecule producing OH and 2H or H_2 (reactions (1) and (2) above), the electrons of the water orbitals will rearrange themselves adiabatically as the H atoms leave so that the unpaired electron that remains on the OH will continue to be in a spatially orientated orbital corresponding to the original chemical bond. The H_2O and the incident H define a plane, and the angular momentum vector of the resulting OH molecule will be perpendicular to that plane; of the orbitals of the OH molecule, some are

Pumping schemes

oriented perpendicular to that angular momentum and others are parallel to it, and for all $^2\Pi_{\frac{3}{2}}$ levels and for $^2\Pi_{\frac{1}{2}}$ levels for which $J \geqslant \frac{9}{2}$, the perpendicular levels have the higher energy (table 5.1). Gwinn and his colleagues conclude that hydroxyl formed from the adiabatic dissociation of water by collision with atomic hydrogen would always be formed in the upper of the Λ-doublet levels of the $^2\Pi_{\frac{3}{2}}$ states and would be formed in the upper levels of the $^2\Pi_{\frac{1}{2}}$ states when J is equal to, or greater than $\frac{9}{2}$.

Table 5.1. *Values of $\langle y^2 - x^2 \rangle$ for various Λ-doublets of OH (Gwinn and others, 1973)*

J	$^2\Pi_{\frac{1}{2}}$		$^2\Pi_{\frac{3}{2}}$	
	l	u	l	u
$\frac{1}{2}$	0.000	0.000		
$\frac{3}{2}$	−0.173	0.172	0.173	−0.172
$\frac{5}{2}$	−0.258	0.256	0.258	−0.256
$\frac{7}{2}$	−0.318	0.316	0.318	−0.316
$\frac{9}{2}$	0.359	−0.361	0.361	−0.359
$\frac{11}{2}$	0.389	−0.392	0.392	−0.389
$\frac{25}{2}$	0.468	−0.470	0.470	−0.468
$\frac{61}{2}$	0.492	−0.494	0.493	−0.493

The average value, $\langle y^2 - x^2 \rangle$, is a measure of the spatial orientation of the wave function; l and u refer to the lower and upper levels of the Λ-doublet.

It is important to note that this mechanism would form CH molecules in the lower levels of the Λ-doublets; the reason is that the p orbitals of OH constitute a nearly complete shell, but with one electron missing, whereas CH has just one p electron; in consequence the orientation of the p orbitals is the opposite of that in OH.

Gwinn and his colleagues also consider the correlation of nuclear spins between H_2O and the OH into which it adiabatically dissociates and conclude that the OH would have the proton spin opposed to the electron spin, that is it would have $F = J - \frac{1}{2}$ in the $^2\Pi_{\frac{3}{2}}$ system. Since the electronic orbitals select the third and fourth levels of a quartet, the nuclear spin correlation picks out the third level. The mechanism however only applies to reactions (1) and (4).

The operation of selection through electron orbitals requires that the water should be in its lowest rotational state, a requirement that limits the temperature of the water.

When OH molecules have formed in upper rotational states they decay rapidly by pure rotational transitions, eventually accumulating in the $^2\Pi_{\frac{3}{2}}$,

5.5 Formation of hydroxyl from water

$J = \frac{3}{2}$ states. Gwinn and his colleagues calculated the transition probabilities for rotational transitions in the cascade down to the lowest level and then calculated the populations of the upper and lower Λ-doublet levels of the $^2\Pi_{\frac{3}{2}}$, $J = \frac{3}{2}$ state, given that the upper Λ-doublet levels of an upper rotational state were preferentially populated. Some results are given in table 5.2 which contains both the relative populations of the Λ-doublet levels of the ground state for each of two reactions and for a range of initial upper states, and also the relative gain coefficients for the respective transitions. It will be seen that reaction (1) strongly favours the 1665 MHz line and reaction (2) the 1667 MHz line.

Table 5.2. *Results of cascade processes for reactions forming hydroxyl (Gwinn and others, 1973)*

(a) Reaction: $H + H_2O + 5.19$ eV $\rightarrow$ OH* + 2H (Text equation, 1)

Starting level	Final relative populations				Relative gains			
	1	2	3	4	1720	1667	1665	1612
25	0.031	0.218	0.397	0.354	0.36	1.48	3.68	0.54
15	0.032	0.216	0.402	0.349	0.36	1.44	3.72	0.55
10	0.032	0.220	0.409	0.339	0.34	1.29	3.79	0.56
5	0.035	0.229	0.426	0.310	0.30	0.88	3.93	0.58

(b) Reaction: $H + H_2O + 0.69$ eV $\rightarrow$ OH + H_2 (Text equation, 2)

Starting level	Final relative populations				Relative gains			
	1	2	3	4	1720	1667	1665	1612
25	0.095	0.155	0.285	0.465	0.37	0.36	1.91	0.39
15	0.095	0.152	0.290	0.463	0.37	3.37	1.96	0.40
10	0.095	0.154	0.293	0.456	0.36	3.28	1.99	0.40
5	0.104	0.156	0.301	0.437	0.32	3.03	1.98	0.41

Levels 1, 2, 3, 4 are designated as in chapter 1. Relative gains are listed under the frequency of the line in MHz. The number under 'starting level' is an index of the rotational level from which the cascade of radiative processes originates; the higher the number, the greater the energy.

The two-fold-processes outlined above – formation of hydroxyl in preferred Λ-doublet levels of high rotational states followed by preferential cascading to the ground state, is somewhat complicated by trapping of infra-red radiation, the consequences of which Gwinn and his colleagues discuss.

The energy required for reaction (1) in particular is quite high and if the hydrogen velocities were thermal, a much higher temperature than the apparent kinetic temperature of the hydroxyl would be required. Further,

Pumping schemes

as mentioned, the rotational temperature of the water must be low. Thus, as in a number of other mechanisms, energetic particles or photons are required for pumping, yet the kinetic temperature of the hydroxyl must be low. It is difficult to find situations where that combination can occur – like others before them, Gwinn and others suggest that the shock front around an expanding H-II region may provide suitable conditions, and they also suggest that a proto-star may do so.

Gwinn and his colleagues argue that their mechanism succeeds in providing a way of inverting populations that favours radiation in the 1667 and 1665 MHz lines of the ground state of hydroxyl. It also predicts the association of hydroxyl with water sources, although as seen in chapter 3, the evidence for that association is not, in detail, convincing. The mechanism is claimed to be consistent with the selection of lines from higher rotational states that are observed. Against those points in its favour, it does not predict inversion of CH Λ-doublet levels.

5.6. Inversion in higher rotational states of hydroxyl

Little attention has so far been given to inversion in higher rotational states, but it is not possible to interpret the observations of maser radiation without understanding how the Λ-doublets of higher states may be inverted, for maser radiation from those states is observed from sources of Turner's class II as well as from his class I. If, for the sake of argument, we suppose that the chemical mechanism of Gwinn and others (section 5.5) operates in the sources of class I, while Litvak's (1969) mechanism of absorption of infra-red radiation operates in the sources of class II, then each of those types of mechanism might be expected to invert populations in the higher rotational states. It can be seen fairly readily that some mechanisms will not invert ground state and excited state populations in the same place. Consider the mechanism of Litvak and others (1966) involving the absorption of ultra-violet radiation (section 5.2). Because the parities of the upper Λ-doublet levels in states $^2\Pi_{\frac{3}{2}}, J = \frac{5}{2}$ and $^2\Pi_{\frac{1}{2}}, J = \frac{1}{2}$ are negative whereas those in the ground state are positive, inversion of the ground state Λ-doublet will be accompanied by anti-inversion of the Λ-doublets of those excited states and by inversion of those of the $^2\Pi_{\frac{3}{2}}, J = \frac{7}{2}$ state. Litvak's far infra-red mechanism could be extended to excited states – for example, the $^2\Pi_{\frac{3}{2}}, J = \frac{5}{2}$ state (from which radiation is observed in NML Cyg) could be populated by radiative transitions from the $^2\Pi_{\frac{3}{2}}, J = \frac{7}{2}$ and $^2\Pi_{\frac{1}{2}}, J = \frac{3}{2}$ states, but of course in competition with decays to and absorptions from the $^2\Pi_{\frac{3}{2}}, J = \frac{3}{2}$ ground state.

5.7. Inversion of the levels in water

A possibly significant difference between the levels involved in the water masers and those involved in the ground state hydroxyl masers is that the former have high rotational energy, they cannot be in thermodynamic equilibrium at low temperatures, and so the numbers of molecules in those levels must depend predominantly on radiative transitions from higher levels.

Two different schemes have been proposed for exciting the water molecules to the higher levels. de Jong (1973) has invoked collisions with hydrogen atoms in a dense cloud of gas, and has argued that the subsequent inversion of the populations in the 6_{16} and 5_{23} levels is a natural consequence of the breakdown of thermal equilibrium in the gas cloud, radiation from different lines escaping from different depths in the cloud. de Jong works out in detail the idea advanced by Jefferies (1971) and already referred to, that thermal equilibrium breaks down and inversions of populations occur in the outer layers of optically thick clouds. de Jong has set up and solved numerically the coupled equations of population rates and transitions in the water molecule and shows that the 6_{16} level will have a greater population than the 5_{23} provided that the decays from the higher states to the lower are controlled by radiative transitions and not by collisions. A possible criticism of de Jong's argument is that it does not take into account the fact that maser radiation is taking place from the 6_{16} to the 5_{23} level and is doing so only from small regions and within narrow cones of directions. Thus the geometrical factors in the equations of transfer differ for the transitions that show maser action and those that do not – the former correspond to restrictions to small regions and narrow ranges of direction, the latter presumably to more or less isotropic radiation.

A different idea has been put forward by Oka (1973). Instead of invoking collisions with hydrogen molecules to excite water molecules into rotationally excited states from which they then decay, as did de Jong, Oka supposes that water absorbs ultra-violet radiation in the band of about 120 to 150 nm. Over that range, as has been shown by J. W. C. Johns in unpublished work at the National Research Council Laboratories in Ottawa, the electronic absorption spectrum of water contains both narrow and diffuse lines, according to the rotational quantum number, those with K_A equal to zero in the upper electronic state lines being much sharper than the others. The other lines are diffuse because of pre-dissociation. The dependence of pre-dissociation on the rotational quantum numbers is well known in other molecules (Herzberg, 1966, p. 458).

Pumping schemes

Now let the levels of the electronic ground state of H_2O be divided into one group (A) with $K_C = J$ and the rest (group B) with K_C less than J. The latter are the ones which will pre-dissociate when they absorb ultra-violet radiation over the appropriate range. The subsequent decay of the excited molecules which have absorbed ultra-violet radiation, but have not dissociated, will be to levels of the electronic ground state of group A. They are the levels of class $J_{0,J}$ and $J_{1,J}$ (the H_2O molecule is rotating almost entirely around the C axis) and when they decay by spontaneous emission without disturbance by collisions, they generally do so to other levels of group A; the exception is to levels of group B with K_C changing by 3, and in particular the 5_{23} level is the first accessible B level lying below the corresponding A level (6_{16}). Thus if the levels of group B are depopulated, the population of level 6_{16} will exceed that of level 5_{23}. Oka points out that there are three processes that could depopulate the levels of group B – molecule formation, pumping and relaxation by collision or relaxation, and dissociation. He argues that dissociation through absorption with ultra-violet radiation will be an effective process. The arguments of de Jong and Jefferies invoke radiation processes to redistribute populations in the outer parts of an optically thick gas and in that way to depopulate the levels of group B.

Oka argues that because ultra-violet radiation is absorbed over a band of some 10 nm in the diffuse lines the necessary intensity of the radiation is less than for mechanisms in which the absorption is into discrete levels over a narrow band.

Oka's mechanism requires an abundant supply of water in the neighbourhood of a hot star. He points out that the water source in W49 has been estimated to radiate some 6×10^{48} photons per second and that at least that number of ultra-violet photons must be produced in a 10 nm band around 110 nm, requiring a star of type O or B in the vicinity. Much of the water will be dissociated by the radiation into OH and H and for the water to be reconstituted there must be a supply of H_2 to enable the reaction

$$OH + H_2 \rightarrow H_2O + H$$

to proceed.

5.8. Some general considerations

Nothing has been said so far of the general orders of magnitude of rates of pumping, but it is possible to set some plausible upper and lower limits. In the first place, remembering that the effective pump rate (chapter 1) is $(p - A)$ where p is the actual rate and A the rate of spontaneous emission,

5.8 Some general considerations

it follows that the actual pump rate must exceed the probability of spontaneous emission for an inversion of population to occur. Upper limits may be estimated for both collisional and radiation processes. The rate of a collisional process can be written as

$$\sigma v N_1$$

where N_1 is the number density of target molecules (OH, H_2O etc.), v is the relative velocity and σ the cross-section for the process which will in general be a function of v, depending on the physics of the postulated pumping process.

In a similar way, the rate of a photon process may be written as proportional to

$$\varepsilon u B,$$

where again u is the radiation density, B is the probability of absorption by target molecules and ε the efficiency of the process. It is assumed here that the absorption step in a chain of radiative processes determines the overall pumping rate.

Consider first, collisional processes. The separation of the O and H nuclei in OH is about 10^{-10} m and the geometrical cross-section of the OH molecule is about 10^{-20} m^2. The geometrical cross-section of an H_2 molecule is of the same order; to an order of magnitude, the cross-section for collision of either H_2 molecules or electrons on OH or H_2O molecules may be taken as 10^{-20} m^2. In a typical gas cloud the density of H_2 molecules may be within one or two orders of magnitude of 10^8 m^{-3}. It is difficult to know what velocity to assume; the equivalent temperature of the molecules may be anywhere from 100 K to 10^4 K, the corresponding velocities lying between 10^3 and 10^4 m s^{-1}. Let us adopt the lower value. The geometrical probability of a collision, $\sigma v N_1$ is then 10^{-9} s^{-1}. No doubt the geometrical cross-section is too great; on the other hand the density of H_2 molecules or electrons may have been underestimated. Thus it seems that collisional pumping, if it is physically realistic, could occur at a rate that exceeds the rate of spontaneous decay in hydroxyl; the rates of spontaneous decay in water, the ground state of CH and excited states of OH, exceed the foregoing estimates of collisional rates by an order of magnitude and it seems less likely that collisional pumping could be effective for transitions in those systems.

Now consider the probabilities of photon-induced transitions. Many possible mechanisms have been suggested and just two examples are considered, both for pumping of OH. The first is pumping by near ultra-violet radiation at 300 nm wavelength, the second by far infra-red radiation at 100 μm wavelength.

Let us assume that the ultra-violet radiation comes from an O or B star with a surface temperature of 2×10^4 K and a radius of 10 times the solar radius, that is 7×10^9 m. The energy density at the surface of the star is then about 10^{-13} J m^{-3} Hz^{-1} and at a distance of 10 A.U. (1.5×10^{12} m) from the star, it would be about 2.10^{-18} J m^{-3} Hz^{-1}.

The spontaneous transition probability for a near ultra-violet transition in OH is about 10^5 s^{-1}, and so the stimulated coefficient B, which is $A(c^3/8\pi h\nu^3)$ is about 5×10^{16} m^3 Hz J^{-1} s^{-1}. Thus the product uB at a distance of 10 A.U. from an O or B star is of the order of 10^{-2} s^{-1}, a value far in excess of the spontaneous transition probability of microwave transitions.

For infra-red pumping, take as source an object of 100 times the solar radius at a temperature of 3000 K. The energy density at the surface is 2×10^{-19} J m^{-3} Hz^{-1} and at 10 A.U. it is 5×10^{-22} J m^{-3} Hz^{-1}. The spontaneous probability for infra-red transitions is of the order of 1 s^{-1} and so B, the coefficient for induced transitions is about 1.5×10^{21} m^3 Hz J^{-1} s^{-1}; the product uB is thus of the order of 1 s^{-1}.

It is evident that the possible probabilities of induced transitions are so very much greater than the probabilities of spontaneous microwave transitions that they impose no significant limitation on pumping processes.

However, these arguments tell us nothing of the total number of photons that emerge from the radio frequency masers as compared with the numbers of ultra-violet or infra-red photons available for pumping. The comparison of probabilities of transitions relates to processes within an elementary volume of hydroxyl gas; the comparison of total numbers of photons relates to the overall gain within the entire maser volume and thus to the geometry of the maser system. It has been mentioned earlier that the overall number of potential pumping photons available whether in the ultra-violet or infra-red seems, so far as estimates can be made (and that is for only a few sources), to be only slightly in excess of the total number of radio frequency photons produced. We therefore seem to arrive at the following position: the probability of collisional pumping processes is no more than comparable with the probability of spontaneous radio frequency transitions, but the number of collisions available is adequate to provide all the photons observed, whereas even though the probability of photon processes is high, the number of pumping photons available is scarcely in excess of the numbers of radio frequency photons generated. We may perhaps conclude that either process may operate

5.9 Creation and destruction of molecules

and that whichever does, the radio frequency power attains its maximum possible value, which seems to be of the same order of magnitude for many sources.

An interesting example of departures from thermodynamic equilibrium occurs in formaldehyde where the upper level of the transition at 1.3 cm is less populated than it would be in thermal equilibrium, thus giving rise to anomalously high absorption. Some of the ideas discussed in this chapter may be drawn upon to explain the phenomenon. Townes and Cheung (1969) have proposed a mechanism relying on absorption of infra-red radiation analogous to that developed by Litvak (1969) for the pumping of hydroxyl levels, but Pelling and ter Haar (1975), on the other hand, have produced a rather convincing model for formaldehyde sources based on the ideas of Jefferies (1971); it is the so-called line leaking model mentioned earlier (p. 92). The same ideas have been employed to explain anomalous intensities in methanol (Pelling, 1975a) and silicon monoxide (Pelling, 1975b).

5.9. Creation and destruction of molecules

Most authors who have studied possible hydroxyl pumping mechanisms have supposed that hydroxyl molecules already exist, Gwinn and others (1973) being a notable exception. But even if it is supposed that hydroxyl molecules already exist before they are pumped, some attention should be given to their formation, particularly in the neighbourhood of H-II regions, the radiation from which is energetic enough to dissociate hydroxyl (and water). Observations of absorption of 18 cm radiation by hydroxyl in cool clouds show that free hydroxyl is not uncommon and may well exist in adequate quantities in the cool clouds around H-II regions and infra-red objects, although the densities in clouds where absorption has been measured are much less than are usually considered necessary for maser action.

Two general schemes are proposed for the formation of molecules in general. The problems that have to be overcome in the formation of molecules are that in order to conserve momentum, three body collisions are required and the probability of such collisions between neutral atoms or molecules in the conditions of interstellar space are too low. Hence it is proposed on one hand that reactions occur on surfaces of grains, from which molecules are subsequently ejected, and on the other hand that ions are involved, for the Coulomb attraction of a charged particle increases its cross-section. The grain hypothesis has been considered by many authors (see for example, Rank, Townes and Welch, 1971; and Williams, 1971); a recent thorough discussion is that of Watson and

Pumping schemes

Salpeter (1972) who show that while molecules will form readily on the surfaces of grains, thermal evaporation is insufficient to remove them from the surface and that ultra-violet radiation is required to do so. The ion hypothesis has recently been considered by Dalgarno (1975).

Apart from separate formation of hydroxyl (or CH or SiO, or H_2O) a plausible idea for the formation of excited hydroxyl from water is that the water is present primarily in grains of ice from which it is sputtered by particles such as protons, hydrogen atoms or hydrogen molecules in the ionization or shock front around an H-II region, and is dissociated almost simultaneously as proposed by Gwinn and others (1973).

The destruction of hydroxyl and other molecules is also important, for the density of molecules where maser action is occurring will be fixed by the balance between the rates of formation and destruction. If hydroxyl is formed or excited in the advancing shock front around an H-II region, it will be dissociated as soon as it is engulfed by the H-II region itself. Thus excited molecules will exist only transiently in the front. Gwinn and others (1973) find that in the presence of H_2 molecules the dominant process leading to the destruction of OH is

$$H_2 + OH \rightarrow H_2O + H.$$

The free water and hydroxyl in the neighbourhood of H-II regions could possibly be produced from ice in the form of grains. If water is present primarily in that form then as the grains enter the ionization shock front of an expanding H-II region, sputtering by high energy protons might both release water from the grain and dissociate it into hydroxyl as in the scheme of Gwinn and his colleagues. Such a mechanism would not however work in the gas around infra-red stars or similar objects, and as with the problem of pumping, different schemes may be required to produce molecules in the neighbourhood of H-II regions and in that of infra-red objects.

6

Analysis and interpretation of maser radiation

6.1. Introduction

Letting it be granted that the intense radiation emitted by hydroxyl and water cannot come from spontaneous transitions, it remains still to account in detail for the observed characteristics of the radiation and to suggest reasonable mechanisms for pumping more molecules into the upper than the lower level of a transition. Much of the published discussion of maser action has concentrated on the pumping problem, partly because it is only recently that sufficient data have become available to enable the structure of a source region to be studied in detail. A somewhat different approach is adopted in this chapter – to see how far the properties of sources can be accounted for by amplification by stimulated emission and to use the analysis to set conditions that must be fulfilled by any pumping mechanism. Other clues to pumping mechanisms may be provided by the statistical properties and associations of maser sources. One aim of such studies would be to use the properties of masers to shed light on conditions in regions thought to be associated with the birth of stars. It must be said at once, that no solutions are given here. Rather, arguments are indicated, and the gaps in our knowledge are shown. Those gaps are of three sorts. In the first place, a few sources have been observed in great detail but it is not known how far the properties of those which have been so observed are typical of general classes of source. Secondly, there are major gaps in the theory of maser processes, in particular a theory of time dependent masers scarcely exists, and the geometry of the sources is uncertain, plausible models notwithstanding. Thirdly, we are ignorant of many properties of the hydroxyl and water molecules which are needed to evaluate the possibility or efficiency of possible pumping processes. At best, it is possible to sketch, in only the broadest way, some possible models of maser sources and processes.

6.2. The properties of maser sources

Apart from the high intensity of the radiation, the properties of maser sources which seem to demand explanation are the small size and

Analysis and interpretation of maser radiation

distribution of the elementary sources, the polarization of the radiation, the emission of radiation from hydroxyl in a single transition, and the variations in time.

It was seen in chapter 4 that masers can be saturated or unsaturated, and it would be very desirable to know whether any particular source is behaving as a saturated or unsaturated maser, for that would give an indication of the rate of pumping. It is usually supposed that the very high intensities imply that most molecular masers are saturated, the intensity increasing in linear proportion to the length of the amplifying column. There is no independent evidence for the supposition, and indeed some evidence against it. The cosmic abundance of deuterium suggests that if the intensity of hydroxyl maser radiation is proportional to the column density of hydroxyl molecules along a line of sight, as in a saturated maser, then radiation from the OD molecule should be observed. It has never been detected, although looked for, and so it seems that the intensity is not simply proportional to the column density of molecules. There are two possibilities. One is that the pumping process is more efficient for OH than for OD; the other is that the intensity depends exponentially and not linearly upon column density, in other words, that the maser is unsaturated. The argument is inconclusive and its strength cannot be tested in the absence of other indications of whether masers are saturated or unsaturated.

6.3. Models of masers

Two ideas already discussed in chapters 3 and 4 are taken as the basis for a possible model of a maser region. The one is that elementary sources lie in a spherical shell at the centre of which is a region of hot gas in an ionized H-II region, or of dust and gas in an infra-red source. If that is so, then the radiation which is amplified by stimulated emission is that emitted by the hot gas or dust at the frequency of the amplifying transition. The question then posed is why the radiation appears to come from a few isolated directions and not from a complete bright rim around the shell. The second idea (mentioned also by de Jong, 1973, among others) is that turbulent motions in clouds of hydroxyl or water molecules will in general so reduce the optical thickness along an arbitrary line of sight, on account of differential Doppler shifts, that amplification will be suppressed, and that it is only along favoured lines of sight that turbulent motions so happen to line up that amplification can occur. It has also been argued that if there are in addition magnetic fields correlated with the turbulent motions, then it is possible to account both for the one sense of polarization shown by hydroxyl radiation and for the amplification of radiation in

6.3 Models of masers

just a single hydroxyl transition. It is almost certain that large scale mass motions take place, with large gradients of velocity, for they are the simplest explanation of the observed spread of frequency in the spectra of hydroxyl and water emissions. It is also very probable that magnetic fields are present, for how else can the radiation be polarized? Thus, rather convincing arguments can be advanced for the constituent elements of the model being discussed; the principal speculative ideas are that it is possible for gas velocities to be sufficiently aligned along *some* lines of sight to permit amplification, and that magnetic fields should be aligned with velocities, as would indeed be the case if the gas were even slightly ionized.

There is very little discussion in the literature of why radiation should be confined to isolated lines of sight (but see Kwan and Thuan, 1974) although it has been pointed out that the radiation amplified by a sphere of gas will be effectively confined to a narrow range of directions through the centre of the sphere. Thus another model that has been proposed for maser source regions involves the concentration of the amplifying gas in a few rather small spheres, each moving with its characteristic velocity. While that model could account for the isolated elementary sources, it says nothing about polarization or amplification in a single transition. No model includes an account of variations in time. On considering the equations of transfer and of population it will be seen that amplification may vary in time because of variations in densities of molecules, because of variations in pumping rates, or lastly, because of changes in the effective probability of stimulated emission corresponding to changes in the degree of alignment of velocities and magnetic fields along the line of sight. A possible clue is that the intensity observed along one line of sight may increase at the same time as that observed along a nearby one is decreasing. It might be thought that changes of alignment of velocities could occur more rapidly from place to place than could changes in densities of molecules or pumping processes.

The close correlation between 18 cm maser radiation and optical and infra-red radiation from late M-type stars suggests that in those sources, it is the pumping mechanism that is varying in time.

The model of amplification along preferred directions with aligned velocities and magnetic fields has been set up to account for the properties of hydroxyl radiation. It could equally account for the isolated directions in which water radiation is seen. Two features of water radiation cannot readily be included if it is assumed that water sources lie in the immediate neighbourhood of hydroxyl sources. In the first place there are instances (W49 for example) where the range of gas velocities corresponding to the spread of frequencies in the water spectra is up to ten times that corresponding to the spread in the hydroxyl spectra.

The polarization of water radiation also presents a problem. Often there is no polarization, but when it does occur it is linear, which on the face of it implies that transverse components of magnetic field are correlated with gas velocity instead of parallel components as proposed for the hydroxyl sources.

The other main differences between water and hydroxyl radiation are the smaller angular sizes of the former (by a factor of 10) the greater brightness temperature (by a factor of up to 100) and the much more rapid variations sometimes shown by water sources. The smaller angular sizes (roughly in the ratio of the wavelengths) might indicate, as has been suggested from time to time, that the angular sizes obtained from interferometric observations do not correspond to the physical cross-sections of the radiating columns but show that there is some lateral coherence in the emitted wave fronts (as in laboratory lasers). Two considerations tell against the suggestion. In the first place it is most improbable that there is any large scale resonant structure in galactic sources corresponding to the optical cavity that imposes lateral coherence in laboratory lasers. Secondly, the apparent sizes of sources of radiation from the higher rotational levels of hydroxyl at a wavelength of 5 cm are much the same as those at the wavelength of 18 cm. Thus, there is no general relation between wavelength and apparent size of source.

No explanation can be suggested for the smaller sizes of water sources. If the angular size is determined by the range of lines of sight over which gas velocities are sufficiently correlated, the angular sizes of water and hydroxyl sources would be expected to be much the same. The greater brightness temperature of water sources is no doubt a consequence of a greater column density of molecules and a more efficient pumping mechanism. As to the faster variations in time, it seems that their origin must be sought in pumping processes or in column densities of molecules, for if the variations arise from changes in alignment of gas velocities they would again be of the same speed as those observed in hydroxyl sources.

To summarize the comparison of water and hydroxyl sources, it seems that a model similar to that for hydroxyl sources, depending on alignment of gas velocities and magnetic field, cannot explain the smaller sizes, the polarization nor the faster variations of the water sources. This may be thought to tell against the model as a plausible one for hydroxyl sources.

There is at present no convincing model that accounts for all the features of maser sources and in particular the apparent angular sizes and

6.4 The two classes of hydroxyl maser

the variations in time do not seem to be explicable in terms of the correlation of gas velocities.

The foregoing analysis of the data does carry some implications for pumping processes. In the first place, it sets the sources in the region between hot gas (H-II region or infra-red star) and the surrounding cool clouds of dust and gas. It suggests that it is in the interaction between hot and cool gas that the processes of excitation and, perhaps, formation, of molecules are to be found. Instabilities in the expansion of the hot into the cool gas (at an ionization front around the H-II region) may account for the variations of intensity from time to time, either because of changes in the column density of molecules or because of changes in the effectiveness of the pumping process.

From a physical standpoint, it is very reasonable to locate the source of radiation in the shell between hot and cold gas. Within the hot gas, especially within an H-II region, molecules would be dissociated; in the cool gas, at a temperature of 100 K or less, there would be no energy to excite molecules. Only where photons or particles escaping from the hot inner zone impinge on surrounding dust or gas can there co-exist both molecules and the particles or photons carrying the energy needed for pumping and perhaps formation.

6.4. The two classes of hydroxyl maser

A recurring theme in this discussion has been whether the causes of the great differences between sources are to be sought in variations of pumping process from place to place and time to time, or in some form of filter, the pumping supposed to be more or less constant. In the foregoing section, attention has been concentrated on the contribution of a filter; now let us turn to evidence for differences in pumping processes.

In Turner's (1970) classification, hydroxyl sources are divided according to whether the main (1667 and 1665 MHz) or the satellite lines (1720 and 1612 MHz) dominate. It is important to bear in mind that sources do not in general radiate *exclusively* in main or satellite lines and it is usually found that sources of class 1 (main line) also radiate at 1720 and 1612 MHz. The significant difference so far as concerns the physics of pumping processes is, as already indicated in chapter 5, that for main line emission to occur the Λ-doublet as a whole must be inverted, that is to say, the population of a magnetic sub-level of a particularly hyperfine level ($F = 1$ or 2) of the upper (+parity) pair of the Λ-doublet must exceed that of a magnetic sub-level of the hyperfine level with the same value of F in the lower (−parity) pair; that is, if n_i is the population of the upper level

Analysis and interpretation of maser radiation

of given F, with degeneracy g_i, equal to $2F+1$

$$n_i/g_i > n_j/g_j,$$

where j refers to the lower level of same F. Labelling the hyperfine levels as in chapters 2 and 5

$$n_4/g_4, n_3/g_3 > n_2/g_2, n_1/g_1.$$

Such a scheme of inequalities allows radiation at 1667 ($4 \to 2$) and 1665 MHz ($3 \to 1$), but it also permits radiation in the satellite lines.

If only one transition is radiating by maser action (($4 \to 2$) is a specific example) then

$$n_4/g_4 > n_2/g_2$$

but

$$n_3/g_3 \leqslant n_1/g_1,$$
$$n_4/g_4 \leqslant n_1/g_1,$$
$$n_3/g_3 \leqslant n_2/g_2.$$

A possible set of populations would be such that n_3/g_3, n_2/g_2, n_1/g_1 are all equal and n_4/g_4 somewhat greater. Then

$$(n_4/g_4 + n_3/g_3) > (n_2/g_2 + n_1/g_1),$$

corresponding to inversion of the Λ-doublet as a whole.

If both main line transitions radiate by maser action then it is clear that again

$$(n_4/g_4 + n_3/g_3) > (n_2/g_2 + n_1/g_1).$$

But now suppose that the 1720 MHz transition ($4 \to 1$) is the only one radiating. Then

$$n_4/g_4 > n_1/g_1$$

but

$$n_4/g_4 \leqslant n_2/g_2,$$
$$n_3/g_3 \leqslant n_2/g_2, n_1/g_1,$$

and a possible set of populations corresponds to

$$n_4/g_4 - n_3/g_3 = n_2/g_2 - n_1/g_1$$

and

$$n_4/g_4 + n_3/g_3 = n_2/g_2 + n_1/g_1.$$

6.4 The two classes of hydroxyl maser

A similar scheme applies for maser radiation in the 1612 MHz ($3 \to 2$) transition. If maser radiation occurs from one or other of the satellite lines, levels with a given value of F (2 for 1720 MHz, 1 for 1612 MHz) have a greater value of n/g than do those with the other value of F, but the Λ-doublet as a whole need not be inverted.

It seems plausible then, that different pumping mechanisms operate in sources of class I and II. In the former, the Λ-doublet as a whole is inverted, and the pumping mechanism does not discriminate between levels with different values of F but in the latter the Λ-doublet as a whole is not inverted but the pumping process does distinguish between different values of F. Furthermore, if the Λ-doublet as a whole is inverted, stimulated emission may occur in satellite transitions as well as in main line transitions, whereas if the Λ-doublet as a whole is not inverted, only one transition and that a satellite transition, can show maser action. The distinction is indeed observed: satellite radiation is often seen from sources of class I but not in general, main line radiation from sources of class II. It seems reasonable then to speak of class I and class II pumping processes as well as of class I and class II sources. Now sources of class I, pumped by processes of class I, are associated with H-II regions, whereas sources of class II, pumped by processes of class II, are associated with infra-red objects. A difference of probable physical significance between H-II regions and infra-red objects is that the photon and particle energies are much greater in the former than in the latter. H-II regions, in which hydrogen is ionized, contain ultra-violet radiation at energies greater than 10 eV, and protons and electrons with about the same energy. The peak energy of the infra-red radiation from infra-red objects is much less. Thus, it is tempting to conclude that high energy particles or photons are needed for pumping processes of class I, whereas photons of lower energy suffice for processes of class II. A word of caution is due here. When Turner drew up his classification, rather simple pictures of H-II regions and infra-red objects suggested that they were quite distinct entities, but it is now clear that there is often an intimate connection between them; in particular, dust may be heated by absorption of ultra-violet radiation (which also ionizes associated hydrogen) and then radiates in the infra-red, so that an H-II region may contain within it infra-red sources. An examination of the physical nature of the objects associated with the two classes of hydroxyl sources is becoming timely. Litvak's (1969) mechanism involving the absorption of far infra-red radiation is of course of class II. Mechanisms of class II are almost certainly photon processes because the only way of distinguishing between levels of the same parity but different F is through the selection rules ($\Delta F = 0, \pm 1$) and probabilities of transitions from higher rotational levels. Although the molecules enter an upper level of the ground state Λ-doublet from outside the Λ-doublet set, none the less the pumping process may be called an internal process, for the molecules excited to the higher rotational levels come originally from the lower levels of the Λ-doublet set.

The scheme of Gwinn and others (1973) inverts the Λ-doublet as a whole and requires energetic particles for collisions with water. It is an 'external' process, for molecules enter the Λ-doublet levels from outside the set, but other processes, effectively of class I, have been proposed, which may be called internal processes, for the molecules entering the upper levels of the Λ-doublet do so from the lower levels.

It is meaningless to classify the pumping processes that operate on water as class I or class II because there is only one upper and one lower level. However, water masers are much more commonly associated with hydroxyl masers of class I than class II so that it may be surmised that high energy photons or particles are needed to pump the water system; Oka's scheme, involving absorption of ultra-violet radiation indeed requires them.

A reasonable overall outline of hydroxyl sources would seem to be that throughout a particular source region the general character of the radiation is determined by the type of pumping process and that the characteristics of the radiation along different lines of sight within the region are determined by the mass velocities and magnetic fields.

6.5. Higher rotational states

For the most part, radiation from higher rotational states of hydroxyl is associated with sources of class I (see section 3) and so it is tempting to suppose that the Λ-doublets of those states are pumped by the energetic processes akin to the class I processes that may operate on the Λ-doublet of the ground state. However, there is one infra-red source which shows radiation from an excited state as well as class II radiation from the ground state.

It is of course evident that molecules that radiate by stimulated emission are not in thermodynamic equilibrium but the presence of significant populations of molecules in higher excited states shows that quite generally molecules are more excited than would be expected at a kinetic temperature of less than 100 K. Not only are there significant populations in levels of the hydroxyl molecule with values of J up to $\frac{9}{2}$ but also water radiation comes from a level with J equal to 6. Maser radiation is one aspect, the most striking no doubt, of a general departure from

6.6 Conclusion

thermodynamic equilibrium of molecules in the neighbourhood of energetic sources.

6.6. Conclusion

The trend of the argument thus far advanced is that at least two pumping processes operate on hydroxyl in the ground state and that possibly more are needed to account for radiation from higher rotational states; and that other evidence – maser action in other molecules, appreciable populations in higher rotational states – witnesses to widespread departures from thermodynamic equilibrium.

Some pumping processes depend in very specific ways on properties of molecular levels, the parity for example, or other factors determining transition probabilities, and in particular, some processes which could invert populations of the ground state quartet of levels of hydroxyl, would not invert those of the next higher rotational state. The facts that, in particular, radiation is observed from the $^2\Pi_{\frac{1}{2}}$, $J=\frac{1}{2}$ state of hydroxyl and from CH imply that general rather than specific pumping mechanisms should be sought if possible.

An analogy may perhaps be drawn with pumping mechanisms effective in laboratory lasers. Laser action can take place in a great variety of gases, in transitions of different types. In certain cases a specific mechanism operates; for example, in the helium–neon laser, the mechanism depends on a near coincidence of energy of a level in neon and one in helium with resonant transfer of energy by collision from helium to neon. More often, however, the inverted populations are brought about because the numbers in each level depend on a balance between radiative transitions and collisions with the walls of the enclosure rather than on the balance with collisions between gas molecules which would produce populations in thermal equilibrium.

In the same way it may be that in celestial masers some very general mechanisms exist that can invert populations of different types of level in a number of molecules.

Let me finally try to see how celestial masers fit into a wider picture. Stars form, so it is believed, from tenuous clouds of interstellar gas, in which by chance condensations develop and then begin to contract under their self-gravitation. As they do so the gravitational energy that is released heats up the gas and dust associated with it, but the pressure of the thermal radiation is insufficient to support the gas against the gravitational attraction until the mass has contracted to roughly stellar size, at which point the temperature and pressure become great enough for nuclear fusion reactions to begin; the extra energy radiated from those reactions then supports the gas against further collapse. Now it is clear that the course of the development of a star from an initial condensation may be affected by what sources of energy other than gravitational energy there may be in the gas, and the sources usually identified are those of magnetic fields and turbulent motion. Further, the course of development may be affected by departures of the gas from thermodynamic equilibrium.

We may ask if observations of celestial masers shed any light on these questions, and whether they suggest others that may be asked about the development of stars? It does indeed seem that they can. First, the velocities equivalent to Doppler shifts of maser radiation from different parts of the gas cloud over a range of about 20 km s^{-1} for sources associated with H-II regions and a greater range, exceeding 100 km s^{-1} in some instances, for sources associated with infra-red objects. In at least one instance the velocities in the second type of source seem to show some systematic rotation; no such clear pattern seems to emerge in sources associated with H-II regions. The tangential velocities of molecules around the margins of H-II regions are comparable with the radial velocities of the shock fronts that bound the regions. If we may take the infra-red objects to be stars in process of formation it seems to follow that velocities in the associated gas are greater before the formation of stars than they are afterwards when the star has reached the O or B stage (if it is massive enough to do so) and a region of ionized gas has developed.

The polarization of hydroxyl radiation shows that there are magnetic fields in the gas wherein stars are forming and that they are of the order of 10^{-7} T or greater. If, as the occurrence of radiation with one sense of polarization at a time suggests, the fields are correlated with the mass velocities of the gas, they would be expected to be larger in the sources associated with infra-red objects than in those associated with H-II regions. Thus, it may be that in the formation of stars the energy of mass motions and magnetic fields is to some extent dissipated.

Finally, consider again the evidence for departures from thermodynamic equilibrium. As has been said before, the very occurrence of maser radiation is itself a major departure, but in addition the occurrence of radiation from rotational states above the ground state is itself evidence that the relative populations of those states exceeds that to be expected for a kinetic temperature of less than 100 K. In general radiation from upper rotational states only comes from the neighbourhood of H-II regions but it cannot in consequence be inferred that the departures from thermodynamic equilibrium are greater in those regions than they are around infra-red objects; the effectiveness of pumping processes

6.6 Conclusion

seems to be greater in the neighbourhood of H-II regions and it may be that even though the higher rotational states are well populated around infra-red objects, none the less pumping between levels within them is not so effective. That consideration apart, it may well be that there are greater departures from equilibrium in the neighbourhood of H-II regions. Such departures are the greater, so it has been argued, the greater the energy of radiation and the lower the density of gas and dust, so that radiational transitions are more likely than collisional transitions. Now it seems probable that the density of matter is greater around infra-red objects (before stars have formed) than around H-II regions (after stars have formed) and at the same time the density of radiation is greater around H-II regions. Greater departures from thermodynamic equilibrium might therefore be expected around H-II regions and that that should be so is consistent with the fact that the more powerful masers are found around H-II regions.

Molecular masers are not the only objects in regions where stars are forming and evidence from other objects must be taken into account in building up a picture of those regions. We cannot so far use molecular masers to determine gas densities for we do not have a complete enough understanding of their behaviour; none the less they do help us to probe the conditions where stars are forming; they show that there are notable departures from thermodynamic equilibrium, mass motions on a large scale and appreciable magnetic fields, and they appear to indicate that the energy of mass motions and magnetic fields is dissipated as stars are formed.

Appendix 1
The Born–Oppenheimer approximation and wave functions for rotating molecules

If the spin of the electron be ignored, the time independent Schrödinger equation for a molecule is

$$\frac{\hbar^2}{2m_e}\sum_i\left(\frac{\partial^2}{\partial x_i^2}+\frac{\partial^2}{\partial y_i^2}+\frac{\partial^2}{\partial z_i^2}\right)\psi + \sum_k\frac{\hbar^2}{2m_k}\left(\frac{\partial^2}{\partial x_k^2}+\frac{\partial^2}{\partial y_k^2}+\frac{\partial^2}{\partial z_k^2}\right)\psi + (E-V)\psi = 0$$

m_e is the mass of the electron and (x_i, y_i, z_i) are the co-ordinates of the electrons, m_k is the mass of a nucleus and (x_k, y_k, z_k) its co-ordinates.

An attempt will be made to solve the equation by finding a wave function ψ which is the product of a function, ψ_e, that depends only on the electronic co-ordinates and of one, ψ_n, that depends only on the nuclear co-ordinates.

Write Δ_1 for

$$\frac{\hbar^2}{2m_e}\sum_i\left(\frac{\partial^2}{\partial x_i^2}+\frac{\partial^2}{\partial y_i^2}+\frac{\partial^2}{\partial z_i^2}\right)$$

and Δ_2 for

$$\sum_k\frac{\hbar^2}{2m_k}\left(\frac{\partial^2}{\partial x_k^2}+\frac{\partial^2}{\partial y_k^2}+\frac{\partial^2}{\partial z_k^2}\right).$$

Schrödinger's equation then reads

$$\Delta_1\psi_e\psi_n + \Delta_2\psi_e\psi_n + (E-V)\psi_e\psi_n = 0$$

or dividing by $\psi_e\psi_n$,

$$\frac{1}{\psi_e}\Delta_1\psi_e + \frac{1}{\psi_n}\Delta_2\psi_n + E - V = 0,$$

provided that $\Delta_1\psi_n$ and $\Delta_2\psi_e$ can be ignored, that is, that quantities like $\partial\psi_e/\partial x_k$, $\partial^2\psi_e/\partial x_k^2$ are negligible. If now V can be split into two parts V_e and V_n where V_e is a function of the electronic co-ordinates and V_n of the nuclear co-ordinates, Schrödinger's equation separates into the two

$$\Delta_1\psi_e + (E_1 - V_e)\psi_e = 0$$

and

$$\Delta_2\psi_n + (E_2 - V_n)\psi_n = 0$$

with $E_1 + E_2 = E$.

The Born–Oppenheimer result is that this separation gives a good first approximation to the wave function of a molecule because the electronic wave function ψ_e

Appendix 1

varies so slowly with the nuclear co-ordinates that quantities such as $\partial \psi_e/\partial x_k$, $\partial^2 \psi_e/\partial x_k^2$, are indeed negligible.

The equation for ψ_n may be shown to be that for a vibrating top. Ignoring the vibration, it becomes

$$\frac{\hbar^2}{2m}\nabla^2\psi + E\psi = 0;$$

the eigenfunctions are surface spherical harmonics,

$$P_J^m(\cos\theta)\, e^{im\phi}$$

where θ and ϕ are the angles specifying the direction of the inter-nuclear axis. m is any of the values $J, (J-1), \ldots, (1-J), -J$.

The eigenvalues are

$$E = \frac{\hbar^2 J(J+1)}{2I}$$

where I is the moment of inertia.

van Vleck (1929) showed that the Born–Oppenheimer separation could still be effected when spin is not ignored. The equation for the rotational wave function then reads

$$\frac{\hbar^2}{2mr^2}\left[\cot\theta\frac{\partial}{\partial\theta} + \frac{\partial^2}{\partial\theta^2} + \frac{1}{\sin^2\theta}\left(\frac{\partial}{\partial\phi} - ij\cos\theta\right)^2\right]\psi + E\psi = 0,$$

where $j\hbar$ is the combined orbital and spin angular momentum of the electrons about the inter-nuclear axis.

The wave functions are Jacobian polynomials:

$$\left[\frac{(d+s+p)!(1+d+s+2p)!}{2\pi p!(d+p)!(s+p)!t^d(1-t)^s}\right]^{\frac{1}{2}} \frac{d^p}{dt^p}[t^{d+p}(1-t)^{p+s}]\, e^{im\phi}.$$

Here m takes the values $J, \ldots, -J$ as before,

$$s = |m+j|, \qquad d = |m-j|$$

and

$$p = J + \tfrac{1}{2}(d+s)$$

while the variable, t, is $\tfrac{1}{2}(1-\cos\theta)$. The eigenvalues are

$$E = \frac{\hbar^2}{2I}[J(J+1) - j^2].$$

Appendix 2
Theories of amplification by tubular and spherical two-level masers

Goldreich and Keeley (1972) set up the maser equations equivalent to those given in chapter 4, in the notation:

$$\frac{dI}{dz} = \frac{h\nu}{4\pi\Delta\nu}[(N_2 - N_1)BI + N_2 A]$$

where B and A are the Einstein coefficients and N_1 and N_2 are the number-densities of molecules in the upper and lower levels.

The rate equations are:

$$\dot{N}_2 = -(N_2 - N_1)BJ - N_2 A + R_2(N - N_{12}) - \Gamma N_2,$$
$$\dot{N}_1 = (N_2 - N_1)BJ + N_2 A + R_1(N - N_{12}) - \Gamma N_1.$$

N is the total number density of molecules involved in the maser, $N_{12} = N_1 - N_2$, R_1 and R_2 are the rates of pumping into levels 1 and 2 (a particular mechanism is postulated here) and

$$J = \frac{1}{4\pi}\int I \, d\Omega.$$

Γ is the decay constant for spontaneous decay. Relaxation processes between maser levels are ignored.

The steady state solution of the rate equations is substituted in the equation of transfer to give the non-dimensional form

$$\frac{d\mathscr{I}}{ds} = \frac{\beta(\mathscr{I} + \frac{1}{2})}{\beta + \mathscr{J}} + S$$

where

$$\mathscr{I} = BI/A,$$
$$\mathscr{J} = BJ/A$$
$$R = R_1 + R_2, \quad \Delta R = R_2 - R_1,$$
$$\alpha = \frac{R}{\Delta R}\frac{A}{\Gamma}, \quad \beta = \tfrac{1}{2}\left(1 + \frac{\Gamma}{A}\right), \quad S = \frac{\alpha\beta}{1 - \alpha},$$

and S is measured in units of the unsaturated growth length, L, equal to

$$\frac{4\pi}{Bh}\left(\frac{\Delta_\nu}{\nu}\right)\frac{1}{\Delta N_0}$$

while $\Delta N_0 = N_{12}/S$ is the unsaturated population inversion.

Appendix 2
1. Cylindrical maser

The maser lies between lengths $-l$ and $+l$.
  The radius of the cylinder is ρ.
  The solutions are given for the amplification of spontaneous emission.

No saturation. The maser is unsaturated throughout if

$$l < l_1,$$

where

$$\tfrac{1}{16}\left(\frac{\rho}{l_1}\right)^2 \exp(2l_1) = 1/\alpha \gg 1.$$

Then

$$\mathscr{I}(l) = (S + \tfrac{1}{2})(e^{2l} - 1),$$

$$\mathscr{I}(l) = \tfrac{1}{4}\left(\frac{\rho}{2l}\right)^2 (S + \tfrac{1}{2})(e^{2l} - 1).$$

Partial saturation. The mean intensity is

$$\mathscr{I}(x) = \tfrac{1}{4}\left(\frac{\rho}{z+l}\right)^2 \mathscr{I}_+ + \tfrac{1}{4}\left(\frac{\rho}{x-l}\right)^2 \mathscr{I}_-$$

where $\mathscr{I}_\pm$ denotes the specific intensity beamed in the directions $\pm x$.
  Note the factors $[\rho/(x \pm l)]^2$ which do not occur in the line maser considered in chapter 4.
  There is an unsaturated core of length $2a$ such that

$$\mathscr{I}(\pm a) = \beta.$$

If $x > a \gg 1$,

$$\mathscr{I}(x) = \mathscr{I}_+(x).$$

Then

$$\mathscr{I}(x) = \tfrac{1}{3}\beta\left[(l+x) + \left(\frac{l+a}{l+x}\right)^2 (3-l-a)\right]$$

for $x > a$, and

$$\mathscr{I}_+(a) = \tfrac{11}{12} Sl^2\, e^{2a},$$

$$\mathscr{I}_+(l) = \tfrac{77}{36} Sl^3\, e^{2a}.$$

a is found to be $\tfrac{1}{2}\ln(\tfrac{48}{11}\alpha\rho^2)$, whence

$$\mathscr{I}_+(l) = \tfrac{28}{3}\frac{Sl^3}{\alpha\rho^2}.$$

2. Spherical masers

Unsaturated sphere of radius R bathed in isotropic background of intensity. In this case, $\mathscr{I}_0 = \mathscr{I}(R, -1)$, that is $\mathscr{I}_0$ is the intensity at radius R, coming from direction

Appendix 2

along the radius vector, i.e., such that

$$\mu = \cos\theta = -1.$$

Then

$$\mathscr{I}(r,\mu) = (\mathscr{I}_0 + S + \tfrac{1}{2})\exp[\{R^2 - r^2(1-\mu^2)\}^{\frac{1}{2}} + r\mu] - (S + \tfrac{1}{2}).$$

$\theta = \arccos\mu$ is the angle between the ray and the radius vector and r is the radius.

$$\mathscr{I}(r) = \tfrac{1}{2}(\mathscr{I}_0 + S + \tfrac{1}{2})\left[\frac{e^R \sinh r}{r} + \frac{(R^2 - r^2)}{2r}\int_{R-r}^{R+r}\frac{e^x\,dx}{x^2}\right] - (S + \tfrac{1}{2}).$$

Partial saturation. The radius of the unsaturated core is given by

$$\mathscr{I}(a) = \beta.$$

For $r > a$

$$\mathscr{I}(r) = \frac{\beta r}{3}\left[1 + \frac{a^2}{r^3}(3-a)\right].$$

Then

$$\mathscr{I}(s) = \mathscr{I}(S_0)\exp\left[\int_{S_0}^{s}\frac{\beta\,dx}{\beta + \mathscr{I}}\right] + S\int_{S_0}^{s}\exp\left[\int_{y}^{s}\frac{\beta\,dx}{\beta + \mathscr{I}}\right]dy.$$

At the boundary of the core

$$\mathscr{I}(a, -1) = \mathscr{I}(R, -1)\frac{R^3}{3a^2} + \frac{SR^4}{12a^2}$$

for $a \ll R$. Since the gain across the core is approximately e^{2a}

$$\mathscr{I}(0, \mu) = \mathscr{I}(a, -1)\,e^a,$$
$$\mathscr{I}(a, 1) = \mathscr{I}(a, -1)\,e^{2a},$$

and so

$$\mathscr{I}(R, 1) = \mathscr{I}(a, -1)\frac{e^{2a}R^3}{3a^2}.$$

Then

$$\mathscr{I}(a) = \tfrac{1}{12}[\mathscr{I}(R, -1) + \tfrac{1}{4}SR]\left(\frac{R}{a}\right)^3 e^{2a}.$$

On putting $\mathscr{I}(a) = \dot{\beta}$ it follows that

$$a^3 e^{-2a} = \frac{\alpha R^4}{48}$$

and the core is just saturated at the centre if

$$4a\,e^{-a} = 1,$$

or

$$a \sim 2.15.$$

Appendix 2

R is then equal to $R_2 = 1.6\alpha^{-\frac{1}{4}}$. If $R > R_2$

$$\mathscr{I}(a) = \frac{SR}{48}\left(\frac{R}{a}\right)^3 e^{2Ka},$$

where

$$\mathscr{I}(a) = \beta/K.$$

and

$$\mathscr{I}(R) = \frac{SR(Ka)}{144}\left(\frac{R}{a}\right)^4 e^{2Ka}.$$

Thus the intensity varies as R^4; the apparent size of the source is $0.7a$ where

$$a = 1.3\alpha^{\frac{1}{4}}R.$$

References

Chapter 1

Cheung, A. C., Rank, D. H., Townes, C. H. and Welch, W. J. 1969. Detection of water in interstellar regions by its microwave radiation. *Nature*, **220**, 62–8.
Cook, A. H. 1966. Suggested mechanism for the anomalous excitation of OH microwave emissions from H-II regions. *Nature*, **210**, 611–12.
Hills, R., Pankonin, V. and Landecker, T. L. 1975. Evidence for maser action in the 1.2 cm transitions of methanol in Orion. *Astron. Astrophys.* **39**, 149–153.
Litvak, M. M., McWhorter, A. L., Meeks, M. L. and Zieger, H. J. 1966. Maser models for interstellar OH microwave emission. *Phys. Rev. Lett.* **17**, 821–6.
Perkins, F., Gold, T. and Salpeter, E. E. 1966. Maser action in interstellar OH. *Astrophys. J.*, **145**, 361–6.
Rank, D. M., Townes, C. H. and Welch, W. J. 1971. Interstellar molecules and dense clouds. *Science*, **174**, 1083–101.
Rydbeck, O. E. H., Ellder, J. and Irvine, W. M. 1973. Radio detection of interstellar CH. *Nature*, **246**, 466–8.
Shklovskii, I. S. 1946. *Astron. Zh.* **26**, 10.
Shklovskii, I. S. 1952. *Astron. Zh.* **29**, 144.
Shklovskii, I. S. 1953. *Dokl. Akad. Nauk. SSSR.* **92**, 25.
Snyder, L. E. 1972. Molecules in Space. Ch. 6 of *Spectroscopy Physical Chemistry Series One* (*MTP Int. Rev. of Sci.* **3**, London, Butterworth).
Snyder, L. E. and Buhl, D. 1974. Detection of possible maser emission near 3.48 millimetres from an unidentified molecular species in Orion. *Astrophys. J.* **189**, L31–3.
Townes, C. H. 1957. IAU Symposium no. 4, p. 92.
Turner, B. E. and Zuckerman, B. 1974. Microwave detection of interstellar CH. *Astrophys. J.* **187**, L59–62.
Weaver, H., Williams, D. R. W., Dieter, Nannilou, H. and Lum, W. T. 1965. Observations of a strong unidentified microwave line and of emission from the OH molecule. *Nature*, **208**, 29–30.
Weaver, H., Dieter, Nannilou, H. and Williams, D. R. W. 1968. Observations of OH emission in W3, NGC 6334, W49, W51, W75 and Ori A. *Astrophys. J. Suppl.* **16**, No. 146, 219–74.
Weinreb, S., Barrett, A. H., Meeks, M. L. and Henry J. C. 1963. Radio observations of OH in the interstellar medium. *Nature*, **200**, 829–31.
Weinreb, S., Meeks, M. L., Carter, J. C., Barrett, A. H. and Rogers, A. E. E. 1965. Observations of polarized OH emission. *Nature*, **208**, 440–1.

Chapter 2

Bluyssen, H., Dymanus, A. and Verhoeven, J. 1969. Hyperfine structure of H_2O and HD Se by beam maser spectroscopy. *Phys. Lett.* **24A**, 482–3.

Chapter 2

Burdyuzha, V. V. and Varshalovich, D. A. 1973. Infra-red and radio transition probabilities of OH and CH. *Soviet Astronomy, A. J.* **16**, 980–2.

Dousmanis, G. C., Sanders, T. M. Jr. and Townes, C. H. 1955. Microwave spectra of the free radicals OH and OD. *Phys. Rev.* **100**, 1735–54.

ter Haar, D. and Pelling, M. 1974. Interstellar hydroxyl, water and formaldehyde masers and dasars. *Rep. Progr. Phys.* **37**, 481–567.

Hall, R. T. and Dowling, J. M. 1967. Pure rotational spectra of water vapour. *J. Chem. Phys.* **47**, 2454–61.

Herzberg, G. 1950. *Spectra of Diatomic Molecules* (New York: van Nostrand Reinhold) 658.

Herzberg, G. 1945. *Molecular Spectra and Molecular Structure – II – Infra-red and Raman spectra of polyatomic molecules* (New York: van Nostrand Reinhold) 632.

de Jong, T. 1973. Water masers in a protostellar gas cloud. *Astron. Astrophys.* **26**, 297–313.

Lichtenstein, M., Derr, V. E. and Gallagher, J. J. 1966. Millimetre wave rotational spectra and the Stark effect of the water molecule. *J. Mol Spectroscopy*, **20**, 391–401.

ter Meulen, J. J. and Dymanus, A. 1972. Beam maser measurements of the ground state transition frequencies of OH. *Astrophys. J.* **172**, L21–3.

Phelps, D. M. and Dalby, F. W. 1966. Experimental determination of the electric dipole moment of the ground electronic state of CH. *Phys. Rev. Lett.* **16**, 3–4.

Powell, F. X. and Lide, D. R. Jr. 1965. Improved measurements of the dipole moment of the hydroxyl radical. *J. Chem. Phys.* **42**, 4201.

Radford, H. E. 1961. Microwave Zeeman effect of free hydroxyl radicals. *Phys. Rev.* **122**, 114–30.

Radford, H. E. 1962. Microwave Zeeman effect of free hydroxyl radicals – $^2\Pi_{\frac{1}{2}}$ levels. *Phys. Rev.* **126**, 1035–45.

van Vleck, J. H. 1929. On σ-type doubling and electron spin in the spectra of diatomic molecules. *Phys. Rev.* **33**, 467–506.

Chapter 3

Baldwin, J. E., Harris, C. S. and Ryle, M. 1973. 5 GHz observations of the infra-red star MWC 349 and the H-II condensation W3(OH). *Nature*, **241**, 38–9.

Ball, J. A., Johnston, K. J., Knowles, S. H. and Moran, J. M. 1972. Interferometer observations of the $^2\Pi_{\frac{3}{2}}$, $J=\frac{5}{2}$ microwave transition. *Bull. Amer. Astronom. Soc.* **4**, 308–9.

Barrett, A. H., Schwartz, P. R. and Waters, J. W. 1971. Detection of methyl alcohol in Orion at a wavelength of ~1 centimetre. *Astrophys. J.* **168**, L101–6.

Baudry, A. 1974. Observations of the excited lines of OH at a wavelength of 6.3 cm. *Astron. Astrophys.* **33**, 381–4.

Buhl, D., Snyder, L. E., Schwartz, P. R. and Barrett, A. H. 1969. An investigation of the spectra and time variation of galactic water sources. *Astrophys. J.* **158**, L97–102.

References

Burke, B. F., Johnston, K. J., Efanov, V. A., Clark, B. G., Kogan, L. R., Kostenko, V. I., Lo, K. Y., Matveenko, L. I., Moiseev, I. G., Moran, J. M., Knowles, S. H., Papa, D. C., Papadopoulos, D. G., Rogers, A. E. E. and Schwartz, P. R. 1972. Observations of maser radio sources with an angular resolution of $0''.002$. *Soviet Astronomy, A. J.* **16**, 379–82.

Chaisson, E. J. and Dickinson, D. F. 1972. OH in the Hoffman infra-red sources. *Astrophys. Lett.*, **12**, 119–22.

Cheung, A. C., Rank, D. M., Townes, C. H. and Welch, W. J. 1969. Detection of water in interstellar regions by its microwave radiation. *Nature*, **221**, 626–8.

Chiu, M. F., Cheung, A. C., Matsakis, D., Townes, C. H. and Cardiasmenos, A. G. 1974. The methanol source in Orion at 12 centimetres. *Astrophys. J.* **187**, L19–21.

Cook, A. H. 1966. Determination of direction, frequency and polarization of radio emission from galactic OH. *Nature*, **211**, 503.

Cook, A. H. 1975. On the structure of hydroxyl maser sources. *Mon. Not. R. Astronom. Soc.* **171**, 605–18.

Cooper, A. J., Davies, R. D. and Booth, R. S. 1971. Interferometric investigations of sources of OH emission. *Mon. Not. R. Astronom. Soc.* **152**, 383–401.

Cudaback, D. D., Read, R. B. and Rougoor, G. W. 1966. Diameters and positions of three sources of 18 cm OH emission. *Phys. Rev. Lett.* **17**, 452–5.

Dickinson, D. F. and Turner, B. E. 1972 Classification of new OH sources. *Astrophys. Lett.* **11**, 1–5.

Evans, N. J. II., Crutcher, R. M. and Wilson, W. J. 1976. Accurate positions of OH emission sources. *Astrophys. J.* **206**, 440–2.

Fillit, R., Gheudin, M., Nguyen-Quang-Rieu, Paschenko, M. and Slysh, V. 1972. New OH sources associated with IR-late type stars. *Astron. Astrophys.* **21**, 317–19.

Gardner, F. F. and Ribes, J. C. 1971. Observations of the excited lines of OH near 4700 MHz. *Astrophys. Lett.* **9**, 175–9.

Gardner, F. F., Ribes, J. C. and Goss, W. M. 1970. Emission of the excited state of OH at 6035 MHz from NGC 6334. *Astrophys. Lett.* **7**, 51–3.

Gardner, F. F., Ribes, J. C. and Sinclair, M. W. 1971. Observations of the $^2\Pi_{\frac{1}{2}}$, $J = \frac{1}{2}$ line of OH at 4660 MHz in Sagittarius B2. *Astrophys. J.* **169**, L109–12.

Goss, W. M., Knowles, S. H., Balister, M., Batchelor, R. A. and Wellington, K. J. 1976. A survey of H_2O sources for wide-spectrum emission. *Mon. Not. R. Astronom. Soc.* **174**, 541–53.

ter Haar, D. and Pelling, M. A. 1974. Interstellar hydroxyl, water and formaldehyde masers and dasars. *Rep. Prog. Phys.* **37**, 481–561.

Harris, Stella, 1974. 5 GHz observations of six OH objects. *Mon. Not. R. Astronom. Soc.* **166**, 29–34 P.

Harvey, P. J., Booth, R. S., Davies, R. D., Whittet, D. C. B. and McLaughlin, W. 1974. Interferometric observations of the structure of main line OH sources. *Mon. Not. R. Astronom. Soc.* **169**, 545–76.

Chapter 3

Hills, R., Pankonin V. and Landecker, T. L. 1975. Evidence for maser action in the 1.2 cm transitions of methanol in Orion. *Astron. Astrophys.* **39**, 149–53.

Johnston, K. J., Knowles, S. H., Sullivan, W. R. III, Moran, J. M., Burke, B. F., Lo, K. Y., Papa, D. C., Papadopoulos, D. G., Schwartz, P. R., Knight, C. A., Shapiro, I. I. and Welch, W. J. 1971. An interferometer map of the water-vapor sources in W49. *Astrophys. J.* **166**, L21–6.

Johnston, K. J., Sloanaker, R. M. and Bologna, J. M. 1973. Thirteen new H_2O sources associated with OH emission in H-II regions. *Astrophys. J.* **182**, 67–75.

de Jong, T. 1973. Water masers in a proto-stellar gas cloud. *Astron. Astrophys.* **26**, 297–313.

Knowles, S. H., Caswell, J. L. and Goss, W. M. 1976. Excited OH radiation from the $^2\Pi_{\frac{3}{2}}, J = \frac{5}{2}$ state in southern H-II regions. *Mon. Not. R. Astronom. Soc.* **175**, 537–55.

Knowles, S. H., Mayer, C. H., Cheung, A. C., Rank, D. M. and Townes, C. H. 1969. Spectra, variability, size and polarization of H_2O microwave emission sources in the galaxy. *Science*, **163**, 1055–7.

Knowles, S. H., Johnston, K. J., Moran, J. M. and Ball, J. A. 1973. Interferometric observations of the $^2\Pi_{\frac{3}{2}}, J = \frac{5}{2}$ state of interstellar OH. *Astrophys. J.* **180**, L117–21.

Litvak, M. M. 1970. Polarized maser emission from interstellar OH and H_2O. *Phys. Rev. A.* **2**, 937–47.

Mader, G. L., Johnston, K. J., Moran, J. M., Knowles, S. H., Mango, S. A., Schwartz, P. R. and Waltman, W. B. 1975. The relative positions of the OH and H_2O masers in W49 N and W3(OH). *Astrophys. J. Lett.* **200**, L111–14.

Masheder, M. R. W., Booth, R. S. and Davies, R. D. 1974. The structure of four 1612 MHz OH emission sources. *Mon. Not. R. Astronom. Soc.* **166**, 561–83.

Meeks, M. L., Carter, J. C., Barrett, A. H., Schwartz, P. R., Waters, J. W. and Brown, W. E. III. 1969. Water vapour observations of galactic maser sources. *Science*, **165**, 180–4.

Moran, J. M., Burke, B. F., Barrett, A. H., Rogers, A. E. E., Ball, J. C., Carter, J. C. and Cudaback, D. D. 1968. The structure of the OH source in W3. *Astrophys. J.* **152**, L97–107.

Moran, J. M., Papadopoulos, G. D., Burke, B. F., Lo, K. Y., Schwartz, P. R., Thacker, D. C., Johnston, K. J., Knowles, S. H., Reisz, A. C. and Shapiro, I. I. 1973. Very long base line interferometric observations of the H_2O sources in W49 N, W3(OH) Orion A and VY Canis Majoris. *Astrophys. J.* **185**, 535–67.

Palmer, P. and Zuckerman, B. 1970. Observations of interstellar OH at 4660 MHz. *Astrophys. J.* **161**, L198–201.

Pelling, Margaret, 1975. A line leaking model for silicon monoxide millimetre wavelength emission. *Mon. Not. R. Astronom. Soc.* **172**, 421–5.

Reisz, A. C., Shapiro, I. I., Moran, J. M., Papadopoulos, G. D., Burke, B. F., Lo, K. Y. and Schwartz, P. R. 1973. W3(OH): Accurate relative positions of water vapor emission lines. *Astrophys. J.* **186**, 537–44.

References

Rickard, L. J., Zuckerman, B. and Palmer, P. 1972. Observations of thermal and maser emission from excited rotational states. *Bull. Amer. Astronom. Soc.* **4**, 307–8.

Rickard, L. J., Zuckerman, B. and Palmer, P. 1973. Detection of thermal absorption and emission by OH in the $^2\Pi_{\frac{3}{2}}$, $J=\frac{5}{2}$ excited state. *Bull. Amer. Astronom. Soc.* **5**, 331.

Rickard, L. J., Zuckerman, B. and Palmer, P. 1975. Observations of quasi-thermal and maser phenomena in rotationally excited OH. *Astrophys. J.* **200**, 6–21.

Rydbeck, O. E. H. and Ellder, J. 1973. Detection of rotationally excited OH in absorption. *Bull. Amer. Soc.* **5**, 477.

Rydbeck, O. E. H., Ellder, J. and Irvine, W. M. 1973. Radio detection of interstellar OH. *Nature*, **246**, 466–8.

Rydbeck, O. E. H., Kolberg, E. and Ellder, J. 1970. OH excited state emissions from W75 B and W3(OH). *Astrophys. J.* **161**, L25–35.

Schwartz, P. R. and Barrett, A. H. 1969. Observations of the $^2\Pi_{\frac{1}{2}}$, $J=\frac{5}{2}$ excited state of OH in W3. *Astrophys. J.* **157**, L109–10.

Schwartz, P. R., Harvey, P. M. and Barrett, A. H., 1974. Time variation of the H_2O maser and infra-red continuum and late-type stars. *Astrophys. J.* **187**, 491–6.

Shklovskii, I. S. 1969. The nature of sources of maser radiation in OH lines. *Soviet Astronomy, A. J.*, **13**, 1–4.

Snyder, L. E. and Buhl, D. 1974. Detection of possible maser emission near 3.48 millimetres from an unidentified molecular species in Orion. *Astrophys. J.* **189**, L131–3.

Sullivan, W. T. III, 1971. Variations in frequency and intensity of 1.35 centimetre H_2O emissions and profiles in galactic H-II regions. *Astrophys. J.* **166**, 321–32.

Thacker, D. L., Wilson, W. J. and Barrett, A. H. 1970. Observations of the $^2\Pi_{\frac{1}{2}}$, $J=\frac{1}{2}$ state of OH. *Astrophys. J.* **161**, L191–7.

Turner, B. E. 1969. A survey for galactic OH emission sources. *Astronom. J.* **74**, 985–93.

Turner, B. E. 1970. Anomalous emission from interstellar hydroxyl and water (parts I and II). *J. R. Astronom. Soc. Canada*, **64**, 221–37 and 282–304.

Turner, B. E., Buhl, D., Churchwell, E. B., Mezger, P. G. and Snyder, L. E. 1970. Observations of interstellar water vapour. *Astron. Astrophys.* **4**, 165–72.

Turner, B. E., Gordon, M. A. and Wrixon, G. T. 1972. Detection of the 4_1–3_0 (E2) line of interstellar methyl alcohol. *Astrophys. J.* **177**, 609–17.

Turner, B. E., Palmer, P. and Zuckerman, B. 1970. Detection of the $^2\Pi_{\frac{3}{2}}$, $J=\frac{7}{2}$ state of interstellar OH at a wavelength of 2.2 cm. *Astrophys. J.* **160**, L125–9.

Turner, B. E. and Zuckerman, B. 1974. Microwave detection of interstellar CH. *Astrophys. J.* **187**, L59–62.

Weaver, H., Dieter, Nannilou, H. and Williams, D. R. W. 1968. Observations of OH emission in W3, NGC 6334, W49, W51, W75 and Ori A. *Astrophys. J. Suppl.* **16**, No. 146, 219–274.

Wilson, A. J., Davies, R. D. and Ellder, J. 1972. The variability of the OH source W3. *Mon. Not. R. Astronom. Soc.* **157**, 21–26P.

Chapter 3

Winnberg, A., Habing, H. J. and Goss, W. M. 1973. Compact radio source associated with the OH source ON-1 (OH 69.5 – 1.0). *Nature (Physical Science)* **243**, 78–81.

Wynn-Williams, G., Weaver, M. W. and Wilson, W. J. 1974. Accurate positions of OH sources. *Astrophys. J.* **187**, 41–4.

Westerhout, G. 1958. A survey of the continuous radiation from the galactic system at a frequency of 1390 Mc/s. *Bull. Astronom. Insts. Neth.* **14**, 215–60.

Yen, J. L., Zuckerman, B., Palmer, P. and Penfield, H. 1969. Detection of the $^2\Pi_{\frac{3}{2}}$, $J=\frac{5}{2}$ state of OH at 5 cm wavelength. *Astrophys. J.* **156**, L27–32.

Yngvesson, K. S., Cardiosmenos, A. G., Shanley, J. F., Rydbeck, O. E. H. and Elldér, J. 1975. Maser radiometer observations of water vapour and OH in weak galactic OH sources. *Astrophys. J.* **195**, 91–9.

Zuckerman, B., Ball, J. A., Dickinson, D. F. and Penfield, H. 1969. Time variations in galactic OH emission sources. *Astrophys. Lett.* **3**, 97–101.

Zuckerman, B. and Palmer, P. 1970. Observations of the $^2\Pi_{\frac{1}{2}}$, $J=\frac{1}{2}$ state of interstellar OH. *Astrophys. J.* **159**, L197–201.

Zuckerman, B., Palmer, P., Penfield, H. and Lilley, A. E. 1968. Detection of microwave radiation from the $^2\Pi_{\frac{1}{2}}$, $J=\frac{1}{2}$ state of OH. *Astrophys. J.* **153**, L69–76.

Zuckerman, B., Turner, B. E., Johnson, D. R., Palmer, P. and Morris, M. 1972. A new interstellar line: the 5_1–4_0 (E2) transition in methyl alcohol. *Astrophys. J.* **177**, 601–7.

Chapter 4

Allen, L. and Peters, G. I. 1972. Amplified spontaneous emission and OH molecules in the interstellar medium. *Nature, Phys. Sci.* **235**, 143–4.

Bender, P. L. 1967. Polarization of cosmic OH 18 cm radiation. *Phys. Rev. Lett.* **18**, 562–4.

Bromley, A. G. 1971. Line interaction in saturated masers. *Proc. Astronom. Soc. Australia*, **2**, 34–6.

Cook, A. H. 1968. Elementary models for stimulated radio emission from interstellar OH. *Mon. Not. R. Astronom. Soc.* **140**, 299–318.

Cook, A. H. 1975. On the structure of hydroxyl maser sources. *Mon. Not. R. Astronom. Soc.* **171**, 605–18.

Deguchi, S. 1974. Radiative stability of the interstellar masers. *Publ. Astronom. Soc. Japan.* **26**, 437–44.

Goldreich, P. and Keeley, D. A. 1972. Astrophysical masers I. Source size and saturation. *Astrophys. J.* **174**, 517–25.

Goldreich, P., Keeley, D. A. and Kwan, J. Y. 1973a. Astrophysical masers II. Polarization properties. *Astrophys. J.* **179**, 111–34.

Goldreich, P., Keeley, D. A. and Kwan, J. Y. 1973b. Astrophysical masers III. Trapped infra-red lines and cross relaxation. *Astrophys. J.* **182**, 55–66.

Goldreich, P. and Kwan, J. Y. 1972. On parametric down-conversion in astrophysical masers. *Astrophys. J.* **176**, 345–51.

Goldreich, P. and Kwan, J. Y. 1974. Astrophysical masers IV. Line widths. *Astrophys. J.* **190**, 27–34.

References

Heer, C. V. 1966. Theory for the polarization of cosmic OH 18 cm radiation. *Phys. Rev. Lett.* **17**, 774–5.

Heer, C. V. and Settles, R. A. 1967a. Theory for the polarization of the four cosmic OH 18 cm lines. *Phys. Lett.* **24A**, 484–5.

Heer, C. V. and Settles, R. A. 1967b. Theory for the anomalous polarization of cosmic OH 18 cm radiation and for hyperfine lasers. *J. Mol. Spectr.* **23**, 448–71.

Lang, R. and Bender, P. L. 1973. Analytical approximation for the saturation behaviour of OH emission regions. *Astrophys. J.* **180**, 647–60.

Litvak, M. M. 1970. Polarized maser emission from interstellar OH and H_2O. *Phys. Rev. A.* **2**, 937–47.

Peters, G. I. and Allen, L. 1972a. Spectral line widths in laser amplifiers and amplified spontaneous emission and its relevance to the interstellar medium. *Phys. Lett.* **39A**, 259–60.

Peters, G. I. and Allen, L. 1972b. Models for OH radiation from the interstellar medium. *Astrophys. J.* **176**, L23–5.

Chapter 5

Cook, A. H. 1966. Suggested mechanism for the anomalous excitation of OH microwave emissions from H-II regions. *Nature*, **210**, 611–12.

Dalgarno, A. 1975. Interstellar molecular absorption lines. *Philos. Trans. Roy. Soc. A*, **279**, 323–9.

Elitzur, M., Goldreich, P. and Scoville, N. 1976a. OH-IR stars I: Physical properties of interstellar envelopes. *Astrophys. J.* **205**, 144–154.

Elitzur, M., Goldreich, P. and Scoville, N. 1976b. OH-IR stars II: A model for the 1612 MHz masers. *Astrophys. J.* **205**, 384–96.

Gwinn, W. D., Turner, B. E., Goss, W. M. and Blackman, G. L. 1973. Excitation of interstellar OH by the collisional dissociation of water. *Astrophys. J.* **179**, 789–813.

ter Haar, D. and Pelling, M. 1974. Interstellar hydroxyl, water and formaldehyde masers and dasars. *Rep. Progr. Phys.* **37**, 481–567.

Herzberg, G. 1966. *Molecular Spectra and Molecular Structure III. Electronic Spectra and Electronic Structure of Polyatomic Molecules* (New York: van Nostrand–Reinhold) 745.

Hughes, V. A. 1967. Mechanism for anomalous OH emission from H-II regions. *Nature*, **215**, 942–3.

Jefferies, J. T. 1971. Population inversion in the outer layer of a radiating gas. *Astronom. Astrophys.* **12**, 351–62.

Johnston, I. D. 1967. A mechanism for maser action of OH molecules in interstellar space. *Astrophys. J.* **150**, 33–45.

Johnston, I. D. 1971. OH emission from proto-stars. *Proc. Astronom. Soc. Australia*, **1**, 336–7.

de Jong, T. 1973. Water masers in a protostellar gas cloud. *Astronom. Astrophys.* **26**, 297–313.

Litvak, M. M. 1968. Interstellar ionised hydrogen (ed. Y. Terzian, New York, Benjamin), 713.

Litvak, M. M. 1969. Infra-red pumping of interstellar OH. *Astrophys. J.* **156**, 471–92.

Chapter 5

Litvak, M. M. and Dickinson, D. F. 1972. OH infra-red stars. *Astrophys. Lett.* **12**, 113–17.

Litvak, M. M., McWhorter, A. L., Meeks, M. L. and Zeiger, H. J. 1966. Maser model for interstellar OH microwave emission. *Phys. Rev. Lett.* **17**, 821–6.

Melrose, D. B. 1973. A plasma hypothesis for anomalous OH emission. *Proc. Astronom. Soc. Australia*, **2**, 206–8.

Oka, T. 1973. *Molecules in the galactic environment* (ed. M. A. Gordon and L. E. Snyder, Wiley Interscience, New York), 258.

Pelling, Margaret A. and ter Haar, D. 1975. Line leaking models for interstellar molecular emission and absorption – I. The anomalous absorption and emission by formaldehyde. *Mon. Not. R. Astronom. Soc.* **171**, 103–18.

Pelling, Margaret A. 1975*a*. Line leaking models for interstellar molecular emission and absorption – II. Masers in the methyl alcohol spectrum. *Mon. Not. R. Astronom. Soc.* **172**, 41–54.

Pelling, Margaret A. 1975*b*. A line leaking model for silicon monoxide millimetre wavelength emission. *Mon. Not. R. Astronom. Soc.* **172**, 421–6.

Rank, D. M., Townes, C. H. and Welch, W. J. 1971. Interstellar molecules and dense clouds. *Science*, **174**, 1083–101.

Solomon, P. M. 1968. Formation and chemical pumping of OH molecules. *Nature*, **217**, 334–6.

Symonds, J. L. 1965. Formation of hydroxyl molecules in interstellar space. *Nature*, **208**, 1195–6.

Townes, C. H. and Cheung, A. C. 1969. A pumping mechanism for anomalous microwave absorption in formaldehyde in interstellar space. *Astrophys. J. Lett.* **157**, L103–8.

Turner, B. E. 1970. Anomalous emission from interstellar hydroxyl and water. *J. R. Astronom. Soc. Can.* **64**, 221–37 and 282–304.

Watson, W. D. and Salpeter, E. E. 1972. Molecule formation on interstellar grains. *Astrophys. J.* **174**, 321–40.

Williams, D. A. 1971. Mechanics of molecule formation. *Observatory*, **91**, 225–7.

Chapter 6

Gwinn, W. D., Turner, B. E., Goss, W. M. and Blackman, G. L. 1973. Excitation of interstellar OH by the collisional dissociation of water. *Astrophys. J.* **179**, 789–813.

Jefferies, J. T. 1971. Population inversion in the outer layers of a radiating gas. *Astron. Astrophys.* **12**, 351–62.

de Jong, T. 1973. Water masers in a protostellar gas cloud. *Astron. Astrophys.* **26**, 297–313.

Kwan, J. and Thuan, T. X. 1974. On the interpretation of the interferometric maps of H_2O Masers near H_2-II regions. *Astrophys. J.* **194**, 293–300.

Litvak, M. M. 1969. Infra-red pumping of interstellar OH. *Astrophys. J.* **156**, 471–92.

Turner, B. E. 1970. Anomalous emission from interstellar hydroxyl and water. *J. R. Astronom. Soc. Canada*, **64**, 221–37 and 282–304.

References

Appendix 1

Herzberg, G. 1950. *Spectra of diatomic molecules* (New York: van Nostrand–Reinhold), 658.

van Vleck, J. H. 1929. On σ-type doubling and electron spin in the spectra of diatomic molecules. *Phys. Rev.* **33**, 467–506.

Appendix 2

Goldreich, P. and Keeley, D. A. 1972. Astrophysical Masers. I: Source size and saturation. *Astrophys. J.* **174**, 517–25.

Index

Absorption, anomalous 92
Aerial temperature 5, 34
Ammonia 2
Angular sizes of masers 31, 34, 80, 82, 88
Atomic hydrogen, radio emission from 1

Born–Oppenheimer approximation 12, 14–16, 118, 119
Brightness, apparent 34, 35, 56, 57
Brightness temperature 5, 34, 35

CH masers 2, 60, 100, 115
CH spectrum 17, 22
Classification of maser sources 111
Collision broadening 35
Collisional pumping 92, 96
Collisions, probability of 103
Coupling of transitions 92

Deuterium oxide 108
Doppler broadening 4, 36, 53
 reduction by maser action 70
Doppler shifts 5, 8, 28, 116

Excited states, formation of hydroxyl in 96

Formaldehyde, anomalous absorption 105

Galactic radio emission 1
Galactic structure 1

H-II regions 2, 9, 36, 37, 38, 42, 54, 56, 91, 105, 110
 excitation of OH 100
 and water masers 60
Hydroxyl 2
 absorption spectrum 5
 destruction 106
 emission spectrum 5
 excited state radiation 55
 far infra-red transitions 94
 formation from water 97
 geometrical cross-section 103
 hyperfine interaction 17
 radiation from higher rotational states 49, 114

rotational levels 14, 15
 Zeeman effect 17–19
Hydroxyl sources and H-II regions 55
 and infra-red stars 55
 associations 53ff, 56
 classification 54
 distribution 53ff
 mass velocities 46, 47
 numbers of components 46, 57
 polarization 8, 44, 46ff, 54
 power 56–8
 variation in time 8
Hydroxyl spectrum 13ff
 intensities in 18–22
Hydroxyl structure 3, 13ff

Ice grains and molecule formation 106
Infra-red absorption and high rotational states 100
Infra-red objects and water masers 60
Infra-red radiation, trapping 99
Infra-red stars 9, 38, 43, 56, 106
Interferometer measurements 31, 36, 38, 39, 42
Interstellar grains 105
Inversion in lambda-doublets, parity effect 100
Inversion of populations 3
Ion–molecule reactions 96, 105, 106

Kinetic temperature 4, 36, 111, 116

Lambda-doublet, inversion of 91
 orientation of molecular orbits 97–8
Lambda-doubling 2, 16ff
Lasers 115
Linear masers 64ff, 69ff
 numerical solutions 77–9, 85–7
Line-leaking 92, 105

M17, excited hydroxyl 50, 53
M-type stars and water masers 60
Magnetic fields 108, 116
Maser equations 70ff
Maser models 108ff
Maser radiation
 isolated directions 36, 42–4, 108, 109
 correlation with other objects 109

133

Index

Masers
  angular sizes 31, 34, 80, 82, 88
  associations 38, 42, 56
  cylindrical 77
  Doppler shifts 28
  equivalent two-level system 82, 83
  general properties 28ff
  instability 83, 85
  kinetic temperature 4, 36, 111, 116
  linear 64ff, 69ff
  location 111
  number of components 36, 42, 46
  numbers of photons radiated 58, 104
  polarization 8, 19, 28, 36, 37, 42, 46, 81, 83ff, 88
  power 91
  quantum mechanical theory 83, 88
  saturated 69, 71ff
  scaling factor 77
  time dependent 32, 33, 47, 85-8, 109
  two level 82
  unsaturated 69
Methanol 3, 60, 105
Molecules
  angular momentum 10ff
  axisymmetric 23
  destruction 105, 106
  diatomic 13
  electronic excitation 12
  formation 105, 106
  polyatomic 22ff
  rotation 2, 11ff
  structure 10ff
  triaxial 23
  vibration 2, 12, 61

NGC 6334 31
  excited hydroxyl 50, 53
  hydroxyl 46
NGC 7538, excited hydroxyl 50, 53
NML Cyg
  excited hydroxyl 50, 100
  hydroxyl 54
  silicon monoxide 61

O Ceti, silicon monoxide 61
OH 69.5 – 1.0, excited hydroxyl 50, 53
ON-1, excited hydroxyl 50
Optical depth 94, 95, 108
Orion A
  excited hydroxyl 50
  hydroxyl 46
  infra-red object 61
  methanol 61
  silicon monoxide 61
  water 34, 58

Parity
  effect on inversion 100
  of lambda doublets 17

Photon absorption, probability 104
Photon numbers 58, 104
Photon pumping 92-5
Polarization 8, 19, 28, 36, 37, 42, 46, 81, 83ff
Polarization, theory 48, 88
Power of masers 56, 58
Pumping 90ff, 102ff
  chemical 67
  classification 91, 113
  collisional 67
  radiative 67, 92-5
  rates 67, 102ff

Radiative transfer
  equation of 64-6
  in spherical geometry 78, 79
  time dependent 66, 85-8
Rate equations 66-7
R Cas, silicon monoxide 61
R Leo, silicon monoxide 61
RX BOO, silicon monoxide 61

Selection rules 93
Sgr B2
  excited hydroxyl 50, 52
  hydroxyl 34, 42, 58
Shock front, H-II region 106
Silicon monoxide 3, 60, 61, 105
Size of maser sources 110
S Per, silicon monoxide 61
Spherical maser 80, 81, 121-3
Spontaneous emission 8, 9
Stars, formation of 115
Statistical weights 63
Stimulated emission 9, 63ff
Super-novae 9

Thermodynamic equilibrium 92, 115
Time dependent masers 8, 28, 29, 32, 33, 47, 57, 85-8
Transition probabilities 19ff, 63, 64
Transitions, selection of 46-9, 108-110
Tubular masers 121
Turbulence 108
Turner's classification 111, 113

U Her, silicon monoxide 61
U Ori, silicon monoxide 61

Very long base line interferometer 38, 39, 57
VX Sgr, silicon monoxide 61
VY CMa
  hydroxyl 34, 38-42, 54, 58
  silicon monoxide 61
  water 34, 58, 60

Index

W3
  excited hydroxyl 50–3
  hydroxyl 34, 46, 47, 54
  objects in region 56
  water 33, 34, 42, 43, 58, 60
W28, hydroxyl 54
W44, hydroxyl 54
W49
  excited hydroxyl 50, 53
  hydroxyl 29, 34, 37, 42, 45, 46, 54, 56, 58, 109
  water 30, 32, 58, 59, 60, 102
W51 7
  excited hydroxyl 50, 53
  hydroxyl 46
W75
  excited hydroxyl 50, 52
  hydroxyl 34, 42, 46, 58

Water
  hyperfine components 27
  inversion of populations 101ff
  level structure 26
  pre-dissociation 101
  pumping by collisions with H_2 101
  spectrum 23ff
  transition probabilities 25–7
Water masers 8, 57ff, 101, 102, 109, 110
  high velocity components 57, 109, 110
  time dependence 31, 59
Wave functions, molecular 118, 119
Westerhout catalogue 38
W Hya, silicon monoxide 61

X Cyg, silicon monoxide 61

Zeeman effect 8, 17ff

Randall Library – UNCW
QB790 .C67
Cook / Celestial masers
NXWW

304900231238Z